THE LAW OF THE MINER

Mining and Society Series

Eric Nystrom, University of Nevada, Reno, *Series Editor*

Our world is a mined world, as the bumper sticker phrase "If it isn't grown, it has to be mined" reminds us. Attempting to understand the material basis of our modern culture requires an understanding of those materials in their raw state and the human effort needed to wrest them from the earth and transform them into goods. Mining thus stands at the center of important historical and contemporary questions about labor, environment, race, culture, and technology, which makes it a fruitful perspective from which to pursue meaningful inquiry at scales from local to global.

Books published in the series examine the effects of mining on society in the broadest sense. The series covers all forms of mining in all places and times, building from existing press strengths in mining in the American West to encompass comparative, transnational, and international topics. By not limiting its geographic scope to a single region or product, the series helps scholars forge connections between mining practices and individual sites, moving toward broader analyses of the global mining industry in its full historical and global context.

Seeing Underground: Maps, Models, and Mining Engineering in America
Eric C. Nystrom

Historical Archaeology in the Cortez Mining District: Under the Nevada Giant
Erich Obermayr and Robert W. McQueen

Mining the Borderlands: Industry, Capital, and the Emergence of Engineers in the Southwest Territories, 1855–1910
Sarah E. M. Grossman

The City That Ate Itself: Butte, Montana and Its Expanding Berkeley Pit
Brian James Leech

One Shot for Gold: Developing a Modern Mine in Northern California
Eleanor Herz Swent

The Weight of Gold: Mining and the Environment in Ontario, Canada, 1909–1929
Mica Jorgenson

Underground Leviathan: Corporate Sovereignty and Mining in the Americas
Israel G. Solares

Smoke and Tailings: An Environmental History of Butte and Anaconda, Montana, 1880–1930
Fredric L. Quivik

The Law of the Miner: The Arizona Experience
John C. Lacy

The Commonwealth of Empire: A History of the Burma Corporation and the Bawdwin Mining Complex in British Colonial Burma
David Baillargeon

THE LAW OF THE MINER

The Arizona Experience

John C. Lacy

UNIVERSITY OF NEVADA PRESS | *Reno & Las Vegas*

University of Nevada Press | Reno, Nevada 89557 USA
www.unpress.nevada.edu

Manufactured in the United States of America

Jacket design: Louise OFarrell
Jacket seal: Seal of the Territory of Arizona, 1863, published by the Office of the *Arizona Miner*, the Official Paper of the Territory, Prescott, Arizona, 1865. The seal was designed by Richard C. McCormick, owner of the *Arizona Miner*. Author's collection.
Jacket illustration: iStock / duncan1890

Library of Congress Cataloging-in-Publication Data
Names: Lacy, John C. author
Title: The law of the miner: the Arizona experience / John C. Lacy.
Other titles: Mining and society series
Description: Reno, Nevada: University of Nevada Press, [2026] |
Series: Mining and society series | Includes bibliographical references and index.
Identifiers: LCCN 2026008398 | ISBN 9781647792503 cloth |
ISBN 9781647792510 ebook
Subjects: LCSH: Mining law—Arizona | Mining districts—Arizona—History |Mineral rights—Arizona—History
Classification: LCC KFA2655 .L33 2026
LC record available at https://lccn.loc.gov/2026008398

The paper used in this book meets the requirements of American National Standard for Information Sciences—Permanence of Paper for Printed Library Materials, ANSI/NISO Z39.48-1992 (R2002).

This book has been reproduced as a digital reprint.

This project is dedicated to the mythical "prudent" miner used by the courts of the United States as a guide for the determination of mineral value. I have never met this person, and I firmly believe that in order to be truly committed to mineral prospecting or mining one must be an optimist far beyond what the "reasonable" person would ever consider prudent.

Myth dies slowly. Sooner than succumb beneath the sharp impact of reality, it tends to assume new forms and live on. . . . Only the name will be changed. Men will know the goal of their desires as El Dorado.

—Stephen Clissold, *The Seven Cities of Cíbola* (1961), 177, 179

Contents

Illustrations

FIGURES

MAP

Preface and Acknowledgments

This project began in 1977 out of an interest in the specific content of the self-governing mining district regulations that formed the "law" for the acquisition of or claim to mineral rights on the public domain of the United States prior to the enactment of the federal mining law in 1866. Much has been written about these practices in California and Nevada during the gold and silver rushes of the 1850s, but very little has focused on Arizona—and knowledge of the specifics of most of the regulations was limited to Clarence King's collection in the 10th Census Report in 1880. My effort was thus to provide as comprehensive collection as could be found of the mining district regulations that grew out of the mineral exploration of Arizona. Once the collection was more or less complete, I needed some framework for a publication, and the initial result was a manuscript far beyond what has been provided here (or, for that matter, publishable). The effort has, however, allowed me to publish a number of articles, my favorites of which include a short note in the newsletter of the Society of Mining Law Antiquarians on the exclusion of lawyers from the mining camps and a history of the development of mining law presented at the 41st Annual Institute of the Rocky Mountain Mineral Law Foundation.[1]

The results of my initial efforts to find the district regulations is contained in appendices A and B and is provided in the hope that it will provide a source for information that is otherwise buried in public records or in sources not easily found. This effort also represents a rather painstaking attempt to decipher obscure handwriting or bad copies. This is particularly true of names, and the author would invite readers of this work to send corrections to the Arizona Historical Society.

Many people have contributed to this final result. Potential readership has always posed a problem as I have attempted to tread carefully on the line between a scholarly contribution that could provide a collection of source materials for historians and students of mining law while at the same time maintaining the general interest of western history enthusiasts. My wife,

Terry, who has both encouraged and endured the research and writing of the manuscript, has suggested for years that titling the work "Shaft!" with a dust cover showing a shirtless, sweating young miner would be its salvation. In the interest of intellectual honesty, however, the concept was, perhaps unwisely, rejected, and this work therefore is destined to fit a somewhat more limited niche.

The contributions of many clients of my legal practice must be acknowledged. They provided me with an opportunity to dig through old records in the course of doing "stand-up" title examinations. These have included, to name just a few, St. Joe Minerals Corporation, Amax Exploration, Bear Creek Mining Company, and Conoco Inc. The individuals who have provided advice and encouragement included especially my wife, Terry; Vic Verity; Stan Dempsey; Bob Pruitt; Don Sherwood; Dave Phillips; Norma Sorrels; Arthur Brant; Trevor Ford; Court Lee; Tony Lane; Tom Reitmeyer; Marsha Luke; Scott Donaldson; Lucille Juliani; and many of my friends associated with the Rocky Mountain Mineral Law Foundation (now the Foundation for Natural Resources and Energy), the Society of Mining Law Antiquarians, and the Arizona Historical Society. I owe a considerable debt for technical assistance to my father, Dr. Willard C. Lacy, whose contributions, combined with the fostering of our mutual love of mining history, has been a genuine source of pleasure. My secretaries and law clerks over the years have also made substantial contributions from deciphering handwritten public records to typing some of the regulations.

THE LAW OF THE MINER

CHAPTER 1

Prologue—Legends and First Attempts at a Mining Law for the New World

This is the story of minerals or, more to the point, the connection of minerals to the creation of wealth and the legal, or simply mutually respected, rights that might be accorded to the person who discovered minerals in the "New World."

In the western United States, the 1848 Treaty of Guadalupe Hidalgo, which ended the war with Mexico, and 1853 Gadsden Treaty, which clarified the southern border, added significant mineral-rich lands to the country. Popular knowledge of the potential value of these lands was triggered by the discovery of gold along the American River in 1848. Because of the timing of the discovery and the date of the Treaty of Guadalupe Hidalgo, coupled with the remoteness of California from the remainder of the United States and the likely inapplicability of the Spanish ordinances, the adventurers of the day were left to their own devices to construct self-governing legal structures memorialized as "mining district regulations." The application of these regulations in Arizona is the subject of this study.

This legal vacuum in California was, however, not the first time that would-be miners faced legal uncertainty. In the 1500s legends of immense riches drew the conquistadors of the Spanish Empire to the New World like moths to a flame. The legends were lent considerable credibility by Spanish conquistadors Hernán Cortés and Francisco Pizarro, who—between the two of them—looted the accumulated wealth of both the Aztecs and the Incas in 1520 and 1533.[1] The Crown of Castile looked upon these expeditions as investments and required each to be specifically chartered. As a part of this process, regal control was exercised through a treasurer appointed to each expedition who was charged with seeing that one-fifth (the *quinto*) of all captured treasure was collected for the Crown.[2]

The legend that provided the catalyst for the mineral exploration of what was to become Arizona was of seven cities, "Cíbola," said to be repositories of great stores of gold. The earliest suggestion of the existence of these cities dates back to a map published in 1505, which places them somewhere in the Atlantic Ocean;[3] however, Cíbola was also reported to be located in the Colorado Plateau of northeastern Arizona. The New World story begins with the 1528 expedition of Pánfilo de Narváez, who held the rights to explore Florida and the Gulf Coast. The expedition was doomed for disaster from its inception; it was poorly equipped and poorly managed, and its relations with Indigenous peoples was poorly handled. The result was that all but four members of the expedition either were killed by Native Americans or died as a result of accidents or starvation. The four survivors included the treasurer of the expedition, Álvar Núñez, who is remembered in history by his ancestral title "Cabeza de Vaca,"[4] along with Andrés Dorantes, Alonso del Castillo, and Estevanico, an Arab slave from Azamor in Morocco owned by Castillo.[5]

During the eight years that followed the destruction of the Narváez expedition, these survivors made their way from the west coast of Florida across the southern United States to Texas, where they followed the Rio Bravo del Norte (Rio Grande) to its headwaters in present-day New Mexico, where they headed southwesterly, possibly crossing the southeastern corner of Arizona before finally making their way to the headquarters of the Spanish authorities on the western coast of Mexico at Mocorito, then on to Culiacán. Once back to civilization, Cabeza de Vaca reported that he knew the location of the cities of Cíbola and that they lay to the north in what is now Arizona. Not surprisingly, the story was of great interest when it reached the highest Spanish official, the representative of the Crown, Viceroy Antonio de Mendoza.

Publicly the story was played down by the Spanish authorities, but quietly Mendoza began a series of efforts to confirm the accuracy of Cabeza de Vaca's story. Under normal circumstances, the word of a man who had been roaming the wilderness for eight years would have seemed suspect, but the persistence of similar stories added to its credibility. When Cabeza de Vaca arrived in Mexico City on July 23, 1536, Viceroy Mendoza attempted to get the members of the party to retrace their steps on his behalf, but all refused. Finally, Mendoza bought Estevanico from Alonso del Castillo and obtained the services of a Franciscan friar, Marcos de Niza, a veteran of both the Mexican and Peruvian conquests, to pursue a mission of verification.

In March 1539, Fray Marcos, accompanied by Estevanico and a small party of support personnel, traveled north through the Santa Cruz Valley of Arizona to the Gila River, then up through Apacheria, the land of the Apache, to northern Arizona. During much of the journey, Estevanico ventured ahead of the group to scout their way; he was apparently killed on one of these excursions, causing Fray Marcos to abort his mission sooner than he intended. In any case, he returned and reported to Viceroy Mendoza that he had indeed seen, although from a distance, the seven cities in the Hopiland of northern Arizona.[6]

This less-than-conclusive report provided little to allay Viceroy Mendoza's concerns about the validity of Cabeza de Vaca's story, and in the fall of 1539 he acted in two definitive ways. First, he dispatched Melchior Díaz, a prominent settler and alcalde at Culiacán, along with Juan de Zaldívar, to check Fray Marcos's story. Obviously, Mendoza found something unconvincing or worrisome in the friar's story.[7] Díaz's expedition went as far north as the Gila River before returning and reporting neither gold nor any evidence that the cities even existed. Apparently, although the news was disappointing, it was still not enough for Mendoza to stop the push for an expedition to Cíbola, perhaps because he was under pressure from others to mount such an expedition, one that he would commission Francisco Vásquez de Coronado to undertake.[8]

On February 22, 1540, Viceroy Mendoza arrived from Mexico City at Coronado's headquarters at Compostela to review the final preparations for the *entrada*. The expedition left the next day.[9] Coronado's authority under Spanish law at this time to claim would have been limited to the disposition of found treasures and would not have given him rights to any mineral deposits found, although, if successful, the leaders of the expedition could expect appointment to administrative posts, supported by a royal salary or the right to collect and use tribute.[10] These possibilities proved to be a moot issue as Coronado found Cíbola to be only the cliff dwellings of the Hopi. Coronado searched further, however, and his journey took him through northern Arizona into New Mexico and as east to what is now the town of Lindsborg in McPherson County, Kansas. The most that Coronado could show on his return were some copper ornaments in the possession of the chief of the Wichitas and an abundance of turquoise jewelry that was used by the Zuñi Indians along the upper reaches of the Rio Bravo.[11] He was also never shown the turquoise mines near modern Cerrillos, New Mexico.[12]

Coronado's journey was considered a disaster by both the Crown and Coronado himself because no riches were found. Neither appreciated the enormity of the exploration effort and wealth of information obtained. The tangible result of the expedition was that the king of Spain, Charles V, was provided with a legal basis to lay claim to the lands in Arizona as far north as the Grand Canyon, but any interest in the mineral wealth of Arizona lay dormant for forty years even as Viceroy Mendoza, through his Viceregal Ordinances of 1550, encouraged the mineral exploration of northern New Spain. The lack of real interest and new mineral *entradas* can be attributed in large part to the success of Spanish mining ventures in central Mexico during the first days of Spanish rule, which operated under the limited rights granted by the Crown and Viceroy Mendoza's ordinances that established legal rights of possession whereby mineral discoveries could be claimed and mined by citizens of Spain.[13]

A significant event in Arizona's mineral exploration came in 1581. By this time the legal structure was again in flux because the *audiencia* of Mexico City had advised the Crown on December 16, 1577, that there were variations between the old Royal Ordinances and the Code of Viceroy Mendoza, and in fact the latter was determinative in lawsuits in New Spain.[14] This lack of authority apparently meant little to Antonio de Espejo, the son of a wealthy miner living in Mexico City, who wanted to search out the potential wealth of Arizona's mineral resources. Espejo's effort was even more tenuous than most of the explorers because all new entries into the northern regions of New Spain could only be done with the express permission of the king. Espejo, however, seized upon an opportunity.[15]

This opportunity arose out of an expedition Franciscan friar Agustín Rodríguez had earlier organized to preach to the inhabitants of present-day New Mexico. His party left Santa Barbara on June 5, 1581, and encountered substantial native hostility. Fray Rodríguez eventually decided to stay at one village and ordered most of the party to return. Because of the hostility encountered on the trip, the Franciscans feared for the safety of those who stayed behind and organized a rescue. It was this rescue mission that Espejo used to justify a mineral *entrada*. Espejo was undoubtedly influenced by reports made by the soldiers on the Rodríguez expedition of the discovery of mine prospects with rich silver and a legend of a lake of gold to the far north.[16] During this mission in March 1583, Espejo found that Friar Rodríguez had been killed, but the stories of rich mines were confirmed while he was visiting the village of Hawikuk near the present site of Albuquerque. Espejo then journeyed west and, on May 8, 1583, apparently found some mineral values in the mountains of central Arizona.[17]

Espejo's own account of his search and discovery of mines is stated as "I found them, and with my own hands I extracted ore from them, said by those who know to be very rich and to contain much silver."[18] The site of Espejo's discovery is not particularly clear, although modern belief is that he crossed the Verde Valley near Mingus Mountain and found the mineral occurrences at Jerome (in which case he would have had to have mistaken copper for silver ore, but early explorers were not above stretching the truth). There is also a chance, however, that Espejo was considerably to the west. In any case, Espejo's own report does not indicate that he attempted to make any official claim on these deposits for himself, but his petitions, first to the viceroy and subsequently to the king after his return indicate that he was seeking some sort of priority or special consideration.[19] Whatever the legalities, however, Espejo's discovery was never worked, the obvious result of the distance to any real civilization.

Official law finally came with the promulgation by King Philip II of the mining ordinances in 1584, which provided a degree of certainty for the acquisition and development of mineral deposits and resulted in a mining boom in central Mexico. Any knowledge of mineral deposits of Arizona, and thus the actual application of any mining law after the promulgation of the 1584 ordinances, does not exist. This was directly related to the absence of an organized government, which in turn would have provided protection from the native peoples who were hostile to the presence of the foreign miners who looked upon them as a source of slave labor. Only the occasional traveler ventured though what was to become Arizona, but any fact-finding mission was usually quickly terminated in favor of safer surroundings of Santa Fe or the various mining centers of central Mexico.

The first step in the establishment of such a government came on April 19, 1583, when King Philip II authorized the award of a contract for the conquest and settlement of the pueblo country along the upper Rio Grande. In doing so, the king was probably influenced by the reports of potential mineral wealth coming from soldiers deserting the fold of Friar Rodríguez and whom Espejo had purportedly attempted to save. The terms of the offer were that the winner had to conquer and settle the lands at his own expense. Little wonder then that the contract was not awarded until September 21, 1595, and the winner was himself rich. He was Juan de Oñate, the son of Cristóbal de Oñate, a miner and former governor of New Galicia who had amassed a fortune from the mining of the silver regions of Zacatecas.[20] On August 18, 1598, Oñate's seat of government at San Gabriel (now known as Ouinge Ouenge Pueblo) near the present site of Santa Fe, New Mexico, established the northern part of what was then known as Pimería Alta and was the first trace of a modern government for Arizona.

In November 1598, the next venture into Arizona took place when Oñate made an expeditionary swing to the west in search of Espejo's silver find. After traveling part of the distance, Oñate commissioned Captain Márcos Farfan de los Godos to take eight men and find the mines. Farfan traveled from the Hopiland of northern Arizona on November 17 through the Chino Valley and along the mountain ranges and valleys of west-central Arizona. Somewhere on either near the Big Sandy River near Wickiup or near the confluence of Kaiser Spring Canyon on Burro Creek, Farfan's men located sixty-six to seventy-two mining claims and returned to San Gabriel with specimens of "rich silver ore."[21]

These claims were located in four separate groups: the first of twenty-eight to thirty claims near an old shaft developed by Native Americans for paint; a second group of fourteen to fifteen claims, which they named the San Francisco vein; fourteen to fifteen claims as the San Gabriel vein; and finally ten to twelve claims as the Guerfanos vein. Under the 1584 ordinances then in force, the discoverer of a new vein could locate as many claims as he wished, and others were limited to two each. Assuming that no particular merit was attached to the discoveries after the first find, and with all the soldiers acting in concert, it is a reasonable supposition that half of the first discovery was reserved for Captain Farfan (and perhaps Governor Oñate), with each of the other eight members of the party taking two claims. The three other discoveries were thus probably the result of finds by groups of five, six or, seven of Farfan's men, each taking two claims. The discoverer's claims would have measured 160 varas (the old Castilian vara was three feet) long and all other claims 120 varas long. Thus, the length of each of the deposits located were approximately 10,528 feet (assuming twenty-eight claims were located, the twelve discoverers' claims and the sixteen regular claims would both be 1,920 varas in length), 4,606 feet, and 3,290 feet, respectively. There is no evidence that Farfan's mines were ever worked or even that the required acts to complete the location were ever followed. Since Farfan returned to San Juan in December, it seems improbable that sixty, or even one or two, nineteen-foot shafts (as required under law) could have been dug prior to their departure.

Among the other claims of mineral discovery in the 1600s were those of Toribio de la Huerta, who promoted a plan to develop allegedly rich silver mines of the Sierra Azul and Cerro Colorado, as well as a mercury mine, between the Zuñis and Hopis. This promotion came to naught when funds from the Crown were not forthcoming.[22]

The next miners in Arizona's history may have followed the Jesuit priests whose trail was blazed by Father Eusebio Francisco Kino. During Father Kino's tenure in Arizona from 1687 until his death in 1711, he was primarily concerned with the advancement of agriculture and stock raising, but his notes do suggest a passing interest in the silver regions accessible from the San Pedro and Santa Cruz Rivers of southern Arizona. The only real evidence of any interest in mining was through Capt. Juan Matheo Manje, Father Kino's military escort during several of his journeys, who reported being told of a silver mine twelve leagues to the west of San Xavier Mission.[23] One enduring legend during Father Kino's era concerns a gold mine supposedly discovered in 1698 approximately one mile south of Oracle, Arizona, in the Santa Catalina Mountains; the mine was supposedly worked for a time by Fray Juan Bautista de Escalante. As with most of these legends, no verifiable evidence of the mine has been discovered,[24] and one suspects that the story was a convenient alibi for a miner accused of high-grading during the boom days of the 1880s.

Perhaps the most significant application of the 1584 ordinances took place in October 1736. At that time, Antonio Siraumea, a Yaqui miner, discovered what appeared to be slabs of native silver southwest of Nogales, Sonora, at a latitude that would have placed it directly south of Tubac. The precise location, near the *ranchería* of Arizona, was also known as "Arizonac." The deposit was soon dubbed the *planchas* or *bolas de plata* (slabs or balls of silver), creating not only the first real rush of Mexican and Spanish miners into the area but also a lasting name for the region, based on the name of the resulting mining camp and the mountain range: Arizona.[25]

The discovery eventually led to litigation that tested the royalty provisions of the 1584 ordinances. The varying reports of the size of the native masses (several pieces were reported in the range of 300 to 500 pounds, with the largest being 2,700 or 3,600 pounds) also attracted the interest of the king's attorney (the *Fiscal*), who saw a chance to shore up the depleted official coffers. Capt. Juan Bautista de Anza (the father of the founder of Tucson and San Francisco) was dispatched from his command at the Presidio of Terrenate to make an official inspection, the assumption being that the mass was a hoard of Aztec silver on its way to the valley of central Mexico. This distinction was important because hoarded treasures were arguably not covered by the new ordinances. A classification of "treasure" would be the key to making it not merely the product of native mining, where only the appropriate royalty would have to be returned to the king. A treasure, on the other hand, could only be obtained through an official charter granted by the king, as was the situation for the early exploration expeditions. Efforts to exercise official control over the deposit finally resulted in an order of the viceroy closing the area in 1741.

The king's investigators eventually determined that the deposit was "natural," but the *Fiscal* determined that the largest mass was a *criadero*, a curiosity, and not covered by the ordinances. Captain de Anza therefore confiscated the piece for the Crown—little wonder that lawyers were not welcome in the mining camps. This decision was appealed, probably by a latecomer because the appeal was not lodged until 1769, to the authorities at Guadalajara. The matter was eventually referred directly to the king who in 1776, not surprisingly, agreed with his attorney.[26] There are no other reported circumstances where a similar position was taken, and the legal treatment of the *planchas de plata* is probably an aberration. It may also have been the realization of the practical impossibility of enforcement. After the authorities seized the find, it was not officially worked, but within a very short time it was gone, and it is likely that the material was fed into a thriving black market for silver to avoid the royalty to the Crown.[27]

Whatever the source of the silver, however, the story of the *planchas de plata* stuck and was frequently repeated. Whether the deposit was natural or a hoard of refined silver made little difference. If the deposit was natural, it represented the possibility that there were more deposits like it. On the other hand, if it was a hoard, it was most likely the result of contraband silver produced from numerous small silver mines throughout what is now southern Arizona. Many of the "Spanish mine" rumors circulated during the early exploration of Arizona were attributed to efforts of the Jesuit missionaries. The Jesuits probably did show some interest in the mineral potential of Pimería Alta, but there is nothing to suggest that this interest was any more than prospecting in the mountain ranges that parallel the Santa Cruz River of southern Arizona. Their presence did, however, act as a stabilizing factor in the incessant "troubles" with indigenous peoples, and it appears that the pueblo of Aribac (later called Arivaca) was the site of the first real settlement that owed its existence at least in part to mining. Stories and evidence of working of the mines at Aribac have persisted, but the actual number of mines has been elusive. A report represented to be of 1777 vintage printed in the *Yuma Sentinel* of April 13, 1878, reported on three mines, with the riches producing ore with a silver content of "one marc per arroba."[28] This specific value is significant and lends a ring of veracity to the report because this value would have to be determined in the process of registering production under Ordinance 22 of the 1584 ordinances. Later, Charles Poston, in his promotion of the mineral potential of his Arivaca grant, acquired in the 1850s, reported that the area around Arivaca contained twenty-five mines that had been conveyed to him by his Mexican grantor.[29] Under the applicable provisions of the code, this would suggest that a nineteen-foot shaft had been dug on each of these mines, and while an examination of the area in modern times would undoubtedly disclose many shafts fitting this description, it seems doubtful that this figure represents registered mines but only prospects generated during perhaps fifty years of active mining operations in the area prior to Poston's arrival. The formality of the mining registration is also called into question by the actual document of conveyance, which makes no specific mention of the mines but simply grants the "mineral rights" along with the property.[30]

The interest in mining did not reach full fruition during the Jesuit control, however. In 1751, the Pima Indians revolted against their converters, and the pueblo was abandoned. De Anza reported, however, that the miners continued to work the deposits until 1767.[31] The Pima revolt was the end of another chapter in the history of Arizona as the Jesuit order was viewed as a growing political force and Charles III of Spain felt threatened. He therefore, in 1767, ordered the expulsion and deportation of members of the Society of Jesus from New Spain. The distrust of religious societies continued beyond this expulsion, and the ability of any religious society to acquire and work the mineral deposits of New Spain was terminated with the new ordinances in 1783.

During the two centuries under the 1584 ordinances, Arizona was in a territorial status called Pimería Alta and loosely attached to either Santa Fe in the north or to what was then the northern frontier of New Spain—Parral. The dispute over the *planchas de plata*, for example, probably started with a local ruling out of Parral and then appealed to the *audiencia* at Guadalajara. Even with peace returning to the Pimas, security for settlers and miners against the Apache continued to be a problem, and in 1776, all of the northern provinces of New Spain were consolidated under a single military authority of the *Provincias Internas* with a headquarters in Chihuahua. The control over Pimería Alta was then with the government of Sonora. Whatever government had jurisdiction over the area of Arizona, however, mining was probably minimal during the period of the existence of the 1584 ordinances.[32] This is suggested by the complete lack of documentation evidencing the registration of mines or the registration and shipment of ores from any mines, which would have been a requirement under the 1584 ordinances. It is possible that the ores from southern Arizona mining, because of its distance from official Spanish control, found its way into the unregistered black market for silver. This seems unlikely, however, as the level of activity that would seem to have been necessary to make such an operation profitable would have most certainly been detected from the logistical support that would have been necessary for the level of production from which ores could be "skimmed."

There were mining activities in what is now New Mexico under the 1584 ordinances near El Paso on the western side of the Rio Grande and at least three claims were filed during the 1750s,[33] but in Arizona, mining activities under the 1783 code were sporadic to say the least. There were several reasons for this. The priesthood, the only permanent residents of the area, other than the indigenous population, were prohibited from the registration and working of mines, and probably more significantly, Arizona was a long way from the supply lines of Mexico, where other richer silver deposits were much more accessible. Because the local governmental structure in Arizona ranged from slack to chaotic, the Apache "depredations" went unchecked.

There was a period, however, beginning during the mid-1780s until approximately 1820, when relative calm settled over the incessant guerrilla war of the Apaches to prevent foreign settlement. This calm was most likely the result of a new policy of providing gifts and food in combination with governmental reform.[34] The isolation of the northern portions of New Spain and the lack of mining activity is evident in the fact that a copy of the code was not transmitted to Santa Fe until almost five years after its promulgation.[35] During this time some mining efforts were able to take a precarious foothold in southern Arizona and continuing along the upper reaches of the Rio Grande, although it is doubtful that the Arizona activities included anything beyond the narrow region of the Santa Cruz Valley.[36] The evidence that does exist includes references to reasonably substantial gold and silver production from Guevavi near Nogales in 1814; to a report that "el Salero," one of the mines in the western foothills of the Santa Rita Mountains, was in operation in 1820; to some gold placers that were being worked in the Sierrita Mountains;[37] and to copper workings that existed at Ajo at the time of its discovery by Anglo trappers in 1836.[38]

The mining activities in New Mexico under the 1783 code brought more permanent mining activities near present-day Silver City, New Mexico. In 1799, retired Mexican colonel José Manuel Carrasco was taken to a native copper deposit by a group of Apaches whom he had assisted while serving as captain of the Presidio of Carrizal in northern Chihuahua. He later formally registered the deposit in 1801, and by June of that year he had registered a number of claims for copper and other minerals and began a history of production at Santa Rita del Cobre that continues today.[39] The population at the mine was found to be 164 individuals in 1806, later grew to between 300 and 400, and continued under Mexican operation until the capture of Santa Fe by United States military forces under Gen. Stephen W. Kearny in 1846.[40]

When the Mexican Republic was established on August 24, 1821, Pimería Alta was recognized as a part of the State of Sonora (which included most of the land within Arizona south of the Gila River), the remainder in a loose territorial status. On February 4, 1824, a law of the National Republican Congress united Sonora and Sinaloa under a single Estado Interno de Occidente that included all of the area south of the Gila. With the collapse of the mission structure in 1828, the presidial system, established by the Spanish and continued under Mexican control, was the only structure of civilization in the Mexican northern frontier. The military establishment of Occidente had its own *comandante general* at Arizpe and had nearly eight hundred officers and soldiers garrisoned at the presidios of Tucson, Tubac, Fronteras, Santa Cruz, Altar, Buenavista, Horcasitas (Pitic), Bocoachi, and Babispe. These garrisons, however, did not have any financial support from the Mexican Republic and could do no more than protect their immediate areas.[41] The Sonorans were never particularly happy with the consolidation (primarily because the problems further to the south did not give sufficient recognition of the need for protection from Apache raids), and Sonora was again resurrected in January 1830. The southern boundary was fixed just south of Alamos, the north and northwestern boundaries being the Gila and Colorado Rivers, and the eastern boundary was the Sierra Madres. The area north of the Gila River was still in question, with the Mexican territories of New Mexico and California both claiming jurisdiction. However, it should be noted that trappers in the area were generally licensed out of Santa Fe, so it is likely that if anyone really cared about the area of Arizona to the north of the Gila River, it was the governor of New Mexico.[42]

One of the few examples of litigation under the Mexican Republic's interpretations of the 1783 code also occurred in present day New Mexico in the gold placers of Los Cerrillos in the Oso (now Ortiz) Mountains. These deposits were discovered in 1828 by José de Jesus Garcia and Rafael Alejo at Sierra del Oro. The deposit was sold on September 18, 1833, to Lt. Francisco Ortiz, who was the military commander at the Real de Dolores, and Ignacio Cano for one hundred pesos.[43] The mining claim itself measured 600 by 102 varas, but Ortiz and Cano also requested grazing and watering places necessary for the operation of the mine and were granted a tract of common pasturing land measuring two leagues in each of the cardinal directions from the mouth of the mine. This grant appears to be the only private land grant for mineral rights eventually approved by the United States under either the Treaty of Guadalupe Hidalgo or the Gadsden Treaty.[44] Part of the asserted legal manipulation related to this mine was that Lieutenant Ortiz, after getting the mine into production, forced out some of his competing colleagues through the use of a decree that forbade any Castilian from residing or operating in New Mexico.[45]

The miners, even with the instability of the Mexican government and the impending war with the United States, seemed oblivious to politics and continued in their efforts to erect a viable legal system. On April 27, 1846, Juan Batista Tournier of Dolores, in a letter to the governor of New Mexico, requested that the territorial tribunal in the area be reestablished. This request was favorably received, and in a resolution of the governor on June 4, 1846, a Provisional Mining Tribunal was established.[46]

Life was not easy in any part of the area, however. Tucson and Tubac were both attacked by the Apaches in the 1830s and 1840s, which forced the withdrawal of settlers, ranchers, and miners. The Apaches were not the sole problem; civil disturbances in Sonora also resulted in extensive destruction of the accumulated archives at Arizpe. Thus, both mining operations ceased, and part of the accumulated history was undoubtedly lost at Arizpe.[47] The names of many of the mines in southern Arizona's Santa Cruz Valley lived on, however, as travelers in 1858 were told of the Salero, Fuller, Encarnation, Bustillo, Crystal, Cazador, and Tenaja Mines, all of which were supposedly worked by Spanish miners.[48]

CHAPTER 2

The Mining Law in the Wilderness of Arizona

The "Kearny Code" of Law

On August 18, 1846, Stephen Watts Kearny, the first commander of the Army of the West, peaceably entered Santa Fe, as Mexican general Manuel Armijo, the governor of the department of New Mexico of the Mexican Republic, fled.[1] Santa Fe was at that time the seat of government for a geographic area including that part of the present state of New Mexico along the Rio Grande Valley from the confluence of the Rio Conejos south to approximately the present United States boundary. The area of present Arizona south of the Gila River was within the Mexican state of Sonora, and the remainder of Arizona was vaguely designated as Apacheria, meaning those lands within the control of the Apache.[2] Kearny's background was in engineering, but he showed a remarkable understanding of what was required to command a civilian population and directed his aide-de-camp, Col. Alexander William Doniphan, to prepare a code of laws. The product of Colonel Doniphan's effort, the Kearny Code and Bill of Rights, was proclaimed law on September 22, 1846, and became the first legal structure of the United States government over the area that would not officially be known as "Arizona" for another eighteen years.[3]

Colonel Doniphan was from a small town in Missouri and had attended Augusta College, a Methodist Episcopal academy, where he graduated with honors in 1826. His legal education consisted of "reading the law," a common practice at the time in lieu of attending law school, and he was admitted to the Kentucky bar by examination in 1829.[4] All of this hardly qualified him to address any aspect of establishing a legal framework for acquisition of mineral rights. Understandably then, the Kearny Code contained no provisions related to mineral rights.[5] In fact, there was little reason to do so; mining was not in fashion, as gold was yet to catch the imagination of the Americans as it had the Conquistadors three centuries earlier. The need might have been foreseen as General Kearny (he had been promoted to brigadier general as a reward for the "peaceful" capture of Sante Fe) had likely visited the copper mines of Santa Rita del Cobre south of Silver City in late 1846, which had recently been worked by both Native Americans and "foreigners" including Sylvester Pattie and James Ohio Pattie during 1825–26.[6]

The Kearny Code adopted the common law of England for all matters concerning real property. However, certain customs in various areas of real property were given a marginal consideration through the following provision: "All laws heretofore in force in this Territory which are not repugnant to, or inconsistent with the Constitution of the United States and the laws thereof, or the statute laws in force for the time being, shall be the rule of action and decision in this Territory."[7] A separate provision was made for the registration of lands and required all persons claiming under Spanish or Mexican grants to file these claims with a register of lands,[8] and special provisions were made concerning water courses and other traditional rights with the provision that "the laws heretofore in force concerning water courses, stock marks, and brands, horses, inclosures, commons, and arbitrations shall continue in force."[9]

The Ad Hoc Mining Rules of the First Prospectors

When gold was discovered in California on January 24, 1848, the treaty ending the war with Mexico had not yet been signed and the only "authority" in California was a small military detachment of the 1st Dragoons of the United States Army commanded by Col. Richard B. Mason. Mason took over as the military governor of California, and one of his first actions was to abolish the laws of Mexico regarding making claim to mines.[10] Later, after a tour of the goldfields, he declined any action to control the miners with a comment that considering "the character of the people engaged and the small scattered force at my commend, I resolved not to interfere, but to permit all to work freely." Colonel Mason did not get any help from Washington either, as a special agent was sent to communicate the views of President Polk that the gold discovery was viewed as transitory and that the government recognized a "great law of necessity." Also, Thomas Hart Benton, the most fervent supporter of Western expansion, cautioned on the floor of the Senate on January 15, 1849, that "this is a case in which the lawgiver must go with the current." The system that evolved in this legal vacuum was a series of consensual self-governing "mining districts" whereby the miners defined the boundaries of a mineral trike, elected officers, and established rules for the locating and maintenance of claims.[11]

While interest centered in California, the mineral potential of Arizona was generally overlooked—and the earlier Spanish and Mexican efforts were lost to history. Although Arizona had been described by John Wesley Powell as being the most "overdrained" area of the world, the drainage came swiftly and in torrents that rarely lent itself to the steady source of water needed for placer mining. It is not surprising, therefore, that several years passed before any significant interest was expressed in the mineral wealth of Arizona. The interest did finally come, however, in the search for legendary Spanish mines.

Trappers and adventure-seekers of the day kept alive the story of "planks" of pure silver in the Mexican state of Sonora, the then-famous *planchas de plata*, "discovered" in 1736. It is likely that Capt. Andrew B. Gray, a member of the boundary commission tasked with surveying the southern boundary of what was then New Mexico Territory, as required by the Treaty of Guadalupe Hidalgo, was the most reliable source of information about the mineral wealth of the area. Gray had heard a great deal about the *planchas* and the "Arrizonia" silver mines during the expedition.[12] Gray sought out the prefect of Altar for information regarding both the *planchas* and the Ajo copper mines, and, based on this information, Peter Brady, a member of his group, set out to examine the sites.[13] The evidence of the *planchas*, meager as it was, turned out to be on the Mexican side of the line, but Brady recognized the oxide ores of Ajo as almost pure copper. In 1854, Brady organized the Arizona Mining and Trading Company in San Francisco with Capt. (Brevet Major) Robert Allen, J. D. Wilson, William Blanding, A. S. Wright, Edward E. Dunbar as manager, and Peter Brady as secretary. The Arizona Mining and Trading Company thus became the first American enterprise engaged in mining within an area that would eventually become Arizona.

The stories of the *planchas de plata* and the copper mines of Ajo were not the only mineralization in "Arizona." Gold placers discovered during the early 1840s in the area of Quitovac in what is now northern Sonora, approximately forty miles south of Ajo, were also reasonably well known.[14] It was clearly the *planchas*, however, that excited the interest of the first American prospectors.

The legal framework for acquisition of mineral rights at the time the Arizona Mining and Trading Company was organized presented a difficult problem. The Kearny Code had been reinforced by its adoption in 1851 by the first New Mexico territorial legislature, but it gave virtually no guidance regarding the acquisition of mineral rights except to create confusion between the adoption of the common law regarding real property and recognizing the laws "not repugnant to, or inconsistent with the Constitution of the United States and the laws thereof." This view was not lost on the general population, and there seems to have been a general sentiment that if the earlier Mexican law regarding the mines was not still legally in force, it should be. For example, the New Mexico convention of delegates, in making a recommendation for a plan of civil government in September of 1849, stated "that the laws of Mexico, heretofore in force, regarding the mineral lands and the working of the mines, be continued in force by making a constitutional provision to that effect."[15] The New Mexican Territorial Legislature reiterated this view on July 7, 1851, and by a joint resolution memorialized the Congress of the United States "that the laws of Mexico on the subject of mines and mining be declared and perpetuated."[16] It is also clear that the legislature did not feel it was operating in a vacuum, as three days later it asked for a geological and mineralogical survey to develop what they considered their vast mineral wealth and artesian wells.[17]

In Ajo in 1854, Brady's group was stymied by the fact that the group was not even sure which side of the border they were on.[18] The miners' customs of self-governing mining districts in California did not seem appropriate since the California practices were developing small, discrete pockets of placer mineralization whereas the Ajo mines were perceived to be large deposits of mineral in-place that did not lend itself to a single individual's effort. Complicating the situation was the language of the treaty for the Gadsden Purchase. Its article VI contained the following provision: "No grants of land within the territory ceded by the first article of this treaty bearing date subsequent to the day twenty-fifth of September when the minister and subscriber to this treaty on the part of the United States proposed to the government of Mexico to terminate the question of boundary, will be considered valid or be recognized by the United States, or will any grants made previously be respected or be considered as obligatory which have not been located and duly recorded in the archives of Mexico."[19] The title to the land was thus encumbered by the possibility of claims under Mexican law. Therefore, the organizers of Brady's Arizona Mining and Trading

Company tried to cover all eventualities and posted papers on their mine claiming the lands under the laws of both the United States and the Republic of Mexico. Formal legal action to acquire the property was initiated in November 1854 when the Arizona Mining and Trading Company took actual possession of the mine. Brady reported that seventeen mining locations were made in 1855 after it was determined which side of the border on which the mine lay.[20]

The place of recording these mining locations is not known, and it is likely that they were not recorded at all. After its acquisition, the area of the Gadsden Purchase was added to the jurisdiction of the New Mexican Territorial County of Doña Ana with its county seat in Mesilla. The entries under Record Book B in Doña County begin on September 26, 1853, and although the first twenty pages are missing, a recordation in 1855 would have been entered in the existing pages.[21]

In the meantime, however, the legal status of the deposit was fought in the pages of the *London Mining Journal*, which was not surprising since the rich ores were being shipped to the smelters in Swansea, Wales. Robert Allen, the president of the Arizona Copper Mining Company, asserted that his company had "denounced" the property under Mexican law. The reports from the *Mining Journal* included the rather simplistic conclusion that "as to titles, the fact of working and occupying mineral lands in the new territories of the United States constitutes the best and most valid claim which hold good in law."[22] The arguments of the United States claimants also stretched to use the homestead and preemption laws to gain title to 160 acres (ignoring of course the legal requirement that the land for these entries were required to be nonmineral in character) while pointing out that the Mexican law would require individual claims.[23] The final title to the Ajo property was not finalized until August 23, 1886, with mineral patents to the Ajo Mining and Development Company represented by W. R. Hunt, and finally on May 22, 1911, when 84.36 acres were patented to Thomas Childs.[24]

The Doña Ana County Official Records

The pattern of locating mining claims within the area of the Gadsden Purchase was clearly reflected in the early records of Doña Ana County, located in Mesilla, New Mexico Territory.

The earliest mining claim recorded in Doña Ana County was the St. Andrews claim of October 5, 1853,[25] by Frank Fletcher and Maj. Enoch Steen. The St. Andrews notice, and those that followed for the next several months, were fairly general and gave no statement of authority. The first serious indication of formality was a petition, using Mexican terminology, for an "order of possession" made to the probate judge of Doña Ana County by Charles Schuchard, Charles Hopkins, and T. G. Miller on March 10, 1854, reciting that "they therefore pray that your honor will order to be spread upon the records of your honorable Court that their claim to said mine which is known by the name of 'Schuchard's Mine' and grant to him all rights and privileges which citizens discovering in opening and working mines are entitled to under and by virtue of the laws of this territory." No specific authority was given, and presumably the locators were talking about the mining laws of Mexico as apparently recognized under the Kearny Code. In any case, the probate judge approved the petition and ordered its recordation.[26]

The format for petitioning recognition of possession was also used by Spanish-speaking miners, and frequently referred to the articles of the *Ordenanzas de Mineria*. For example, the "Trinidad" mine in the Organ Mountains requested possession under the authority of book 6, article 10, which under the 1783 *Ordenanzas* gave authority to take possession of abandoned claims.[27]

The format for future petitions was cast in a petition made on July 18, 1854, for an unnamed mine in the Soledad Mountains, which was filed by Marriano Basela, Alejandro Daguerre, and Thomas J. Bull, who made the following statement in their petition: "We would register the above mine according to the Mexican Mining Law, which we believe to be in full force in this territory, but should it at any time appear that said laws are not in force, then in that case we would wish to hold the same under the pre-emption laws of the United States."[28] This same petition was repeated many times throughout old records.

There was also some recognition of a requirement of diligence in working of the claim, and in an undated notice filed by Hugh Stephensen and James A. Lucas, it was asserted that the "Santa Adila" claim had been "altogether neglected and abandoned" and that the petitioners had taken possession of the claim. The notice requested official recognition of the forfeiture of the older title and to note on the record that the claim had been renamed as the "Congress Mine."[29]

The size of these claims and requests for orders of possession are generally not stated, and presumably the locator was simply marking the extent of his possession by monuments on the ground. One claim that does include some dimension was located on April 1, 1858, by William Claude Jones, George W. Southwick, and Caleb Sherman, who located the Puerto de San Augustine in the Organ Mountains of New Mexico. The notice included a plat showing a claim approximately 3,000 yards in length by 300 yards in width.[30]

There are eight mines within the Doña Ana County records that are clearly within that portion of the Gadsden Purchase that would ultimately be included within the State of Arizona. The first two mines were the French Mine and the Crop Mine located seventeen miles north of Santa Cruz, Sonora, and eighteen miles south of Fort Buchanan. The claims were located by H. T. Titus, W. J. Godroz, George D. Mercer, Elias Brevoort, and B. J. D. Irwin on April 25, 1859. The petition for this mine requested "that same may be registered to their sole use and benefit in accordance with the 'Ordinances de Mineria' of Mexico now in force in said territory, and they pray that this their petition may be allowed, approved and registered amongst the records of said county for future observance, and that the corresponding certificate may issue."[31] The probate judge allowed and approved the petition and ordered the claim to be registered.

The next petition to be considered by the probate judge and "Prefect," the Honorable Rafael Ramlas, was for the "Hawley Mine" and was submitted on behalf of Giles Hawley, Samuel W. Cozzens, Frank D. Ryther, and Edward Dickenson, who asserted that they discovered a vein of silver in the Dragoon Spring Mountains to the south of the overland mail route. Their petition stated: "They are desirous that the same should be denounced according to the Ordenanzas de Mineria of Spain and Mexico which are still in force in this territory." The petitioners asked to be placed in legal possession of a claim that was 4,000 yards square and that had the corners monumented. They also petitioned that the claim include "the minerals, timber, wood, water, rock and materials necessary for working said mine." The petition was allowed on July 20, 1859.[32]

Probably the best documented mine was that of the Maricopa Mine (also known as "Gray's Mine") originally located as the "Aztec," but since the name was already in use by the Lake Superior Copper Company, the name was changed. The first official record of the claim was a relocation notice changing the name of the claim submitted by Charles C. Walters "residing in the Arizona Territory of New Mexico."[33] The claim was described as thirty to forty miles easterly from the "Sacation [*sic*] Station of the Overland Mail Company now called Pimo Station" near Paso Verde. The notice recited that Walters, along with Arizona boundary surveyor

Andrew B. Gray, discovered a vein of metallic ore on December 26, 1859. Walters described his discovery as copper carbonate and stated that it had been worked by J. George Bryant, a "practical miner" who also worked for Gray as a field assistant. The notice recited a discovery date of December 26, 1859, but was not filed until March 4, 1860, along with a similar notice by Bryant.[34] Gray also described the find in a separate statement that he "located the same with the intention of complying with all the laws appertaining to mining or preemption rights."[35]

The claim was a quarter section or 880 yards square and was described in relation to a monument that Gray took pains to describe with reference to a point he fixed as 33 degrees 53 minutes north latitude and 10 degrees 50 minutes west longitude. The recording of the longitude is clearly erroneous, as it is known to be ground relocated as Eureka number 4 patented claim of William Mill Williams.[36] But if it is 110 degrees 50 minutes, and if the longitude were 33 degrees 35 minutes, the discovery might have been the Pinto Creek copper deposit near Miami, Arizona.

The Doña Ana County records also indicate that Gray was somewhat of a promoter of his mine and sold fifty shares of stock in any company to be formed concerning the mine to the Honorable P. T. Herbert and Granville Oury.[37] The final indication of proceedings concerning the Maricopa Mine was a Quitclaim Deed dated July 7, 1860, whereby Gray, Robert South, James J. Lautter, Frederick H. Wolcott, Charles W. Wolcott, and Douglas Robinson deeded their rights to the Maricopa Mining Company.[38] The last Doña Ana County record of Gray's work in Arizona is an 1861 lawsuit in Mesilla, at the then Confederate Court, Territory of Arizona, CSA, over debts for survey work. By then Gray had enlisted in the Confederate army engineers and in April 1862 was killed at Fort Pillow, Tennessee.[39]

The final Arizona entry in the Doña Ana County records was the location of the Lucero, Aurora, Encanto, and Delecias silver mines, which were described as eight to ten leagues from the Arivaca Ranch in the direction of the "Papagoes Country."[40] The mines were located on April 14, 1860, by Jose Fedirico, Manuel Sola, Juan Jose Salgado, Francisco Campillo, Nacario Ortiz, Damian Radereco, and Thomas Smith.[41]

The Doña Ana records during this time also suggest that a separate attempt to create a records repository in Tubac by Charles Poston (see next section) was suspect as Poston recorded his lease for the "Arrivacca Ranche" and the Heintzelman Mine in those records on October 16, 1860,[42] and Sylvester Mowry recorded his purchase of the Patagonia Mine (now in Santa Cruz County) in the same records. This purchase was made on April 9, 1860, for $22,500 from Henry P. Titus, Mary E. Titus, and Elias Brevoort,[43] although Mowry has in other sources indicated that the purchase had been made from 1st Lt. Isaiah N. Moore (the commander of Fort Breckinridge), 2nd Lt. Richard S. C. Lord, and Richard S. Ewell, along with James W. Douglass and Richard Doss.[44]

As the Civil War began to disrupt governmental authority, the political boundaries began to fade, and the Doña Ana County records began to make references to the Territory of Arizona, County of Doña Ana, the Eastern or Western District of Arizona, or "Alto Arizona."[45] And then from August 1861 to June 1862, Doña Ana County, Arizona Territory, Confederate States of America" also appeared in the records.[46] In fact, the period from 1860 to 1863 adds considerable support for Poston's efforts to control his own records, as the seat of "local" government was a moving target during this period. The uncertainty began with an act of February 1, 1860; by section 1, the New Mexico Legislative Assembly created the County of Arizona described as lands situated west of a north-south line established at one mile east of the Overland Mail Station to Apache Canyon. Tubac was designated the county seat. By an act of January 8, 1861, the county seat for Arizona County was changed from Tubac to Tucson. The county itself was short-lived, however, as by an act of January 18, 1862, the law establishing Arizona County was repealed, and the area again included in Doña Ana County, only to be overturned again on January 28, 1863, when Arizona County, as originally created by the 1860 act, was again "put in force in all their vigor."[47]

The Alcalde of Tubac

Unsurprisingly, the political uncertainty invited individual accommodation. In early 1854 Charles D. Poston and Herman Ehrenberg made a brief visit to what would become Arizona in a search for the *planchas de plata*. Poston was born near Elizabethtown, Kentucky, on April 20, 1825, and at an early age served as a clerk in the Harden County Courthouse. He read law and performed duties as an attorney as a young man. In 1850, after California became a state, he obtained a job in a customs house in San Francisco and probably had political ambitions within the new state. Herman Ehrenberg was a native of Germany who had been schooled as a mining engineer. At eighteen, Ehrenberg had fought for the Republic of Texas, surviving the Fannin Massacre at Goliad, and later took part in the Battle of San Jacinto before returning to Germany to complete his training. He then returned

Fig. 1. Charles D. Poston (1825–1902). The self-styled "Alcalde of Tubac" who promoted and developed silver mines in the Cerro Colorado Mining District from a headquarters in Tubac, Poston was known as the "Father of Arizona" for his lobbying for the creation of Arizona Territory. Courtesy of the Arizona Historical Society, 60594.

to Texas, journeyed to Hawaii and elsewhere in the South Seas, and served under John Charles Frémont during the Bear Flag War in California.

Poston, along with Ehrenberg, while in California, formed the Sonora Exploring & Mining Company and purchased the unconfirmed Mexican Arivaca Land Grant, which recited that it included twenty-five "mines" (almost certainly no more than prospects). This purchase was probably motivated by the unsettled state of mineral titles and the general concern of the day that the rights within these grants might be insecure and wanting to cover all bases.[48] Poston also took steps to establish himself as the primary legal authority within southern Arizona, and to this end on July 15, 1856, obtained an appointment as deputy clerk for Doña Ana County of New Mexico Territory.[49] This "authority" made him the only representative of civil government in the area except for the occasional visit of one of the federal judges of the Territory of New Mexico. He styled himself as an "alcalde" and, based on this authority, claimed to have frequently performed marriages, which eventually led to some difficulties with the Roman Catholic Church.[50]

Poston also acted as the local recorder, and his record books are maintained in the records of the Pima County Recorder in Tucson as "Old Records Book No. A." By this record, Poston himself set the form for initiation of mineral rights by locating the first mining claim on February 1, 1857. Poston called the claim the "Salero,"[51] a purported relocation of the earlier Spanish- or Mexican-era mining claim of the same name and recited that it was claimed "under the laws of Mexico or of the United States" apparently still out of uncertainty as to which sovereign's laws would eventually control the mines. It is also not surprising that he waited until 1857 to initiate the claim as it was not until March 10, 1856, that the Mexican commander left Tucson—and the troops of the United States did not arrive until November. Poston, during this period, was in California organizing the Sonora Exploring & Mining Company, and the location of the Salero was probably one of his first acts upon his return. Ironically, the ground chosen by Poston for this first location was officially considered "non-mineral in character" only nine years later when application was made for selection of the Baca Float No. 3 private land grant.[52] The Salero's location notice indicates the claim was situated in the Santa Rita Mountains and was marked to a depth of eighty feet.[53] The parties indicated as having an interest in the claim included Frederic Brunckow, Charles Schuchard, Herman Ehrenberg, Theodore Moohrman, George W. Fuller, Samuel N. Heintzelman, Edgar Conkling, and William Wrightson. Similar location notices were also of record for the Heintzelman Mine, located by Brunckow, and the Aranea Mine, located by Schuchard, on February 1, 1857.[54]

The common depth for the sinking of shafts on these claims of Poston and his associates was thirty feet, paralleling a requirement under Mexican law that required a location shaft of ten varas deep.[55] Knowledge of the laws of Mexico, and Mexican mining law in particular, appears to have been reasonably widespread. This might be attributed to the fact that the law had been in effect for almost seventy-five years when the first "American" mining claims were located. The availability of Spanish legal books likely also played a role. Samuel Heintzelman, recently retired as an army general, who had been recruited by Poston to manage the operations of the Sonora Exploring & Mining Company, provided detailed insight through a diary he kept between August 3, 1858, and January 29, 1859, noting that on December 26, 1859, he had finished reading Francisco Xavier Gamboa's classic 1761 treatise on the Spanish mining law.[56]

Heintzelman's perspective on life in Tubac in the late 1850s was not quite the picture of paradise that Poston is generally credited with painting, and Heintzelman's descriptions included discussions of both criminal and mining law. The criminal aspects involved high-grading at the mine, which was overlooked upon the first offense, but fifty lashes were promised for the next occurrence. Later, when a murder was committed at the mine, the perpetrator would have likely been executed except that he escaped.[57] Life in and around the mining operations was clearly hazardous. Persons traveling alone or without armed protection were frequently killed by Apaches, and in 1861 the entire management at the Santa Rita Mine was murdered by Mexican bandits with the tacit approval or indirect assistance of the mine's labor force (who were probably by and large also escapees from the Mexican judicial system).[58]

The records suggest that during the early life of these mining operations, there may have been some concern about the ability of a business entity to hold title to the lands as the legal title to the claims remained in the names of individuals. Poston had formed his Sonora Exploring & Mining Company shortly after his arrival in the area, but it was not until October 15, 1860, that the claims were formally leased to the company.[59]

In spite of the "official" nature of both Poston's actual position of deputy clerk of Doña Ana County and his claimed office of alcalde of Tubac, Poston, his associates, and the Sonora Exploring & Mining Company were apparently the only ones who recognized the official nature of Poston's records. This seems evident because the location notices of Poston and his associates are the only mining location notices recorded within the book. The method of locating and preserving title to the mines did, however, follow a reasonably uniform pattern throughout the area.

The First Gold Rushes

The gold rushes that descended along the Colorado River and into the Gila River tributaries from 1858 to 1863 were a small part of a much larger historical drama. The California Gold Rush of 1848–49 had set off a pattern of mineral discovery that brought a large population to the western slope of the Sierra Nevada. For a decade California established a pattern for placer mining based on Spanish-American and European traditions, in styles and technology, governmental, law, and social life—which was imitated by subsequent gold rushes into the western territories. With the decline in production of the Mother Lode, the discouraged California miners brought these systems first to Nevada and Colorado (1859), then Idaho (1862) and Montana (1863), and to Arizona's placer mining rushes of the 1850s and 1860s.[60]

Like the earlier tales of cities of gold and of *planchas de plata*, the gold rush era was also filled with tales of mythic wealth to be had. Lieutenant Colonel Emory's report on the 1846 mapping of the Gila River is sprinkled with the word "gold." He wrote that the Rio Prieto tributary "flows down from the mountains, freighted with gold. Its sands are said to be full of this precious metal," a description tempting prospecting parties for more than two decades—the Capt. William Love party of 1852, the Joseph R. Walker party in 1863, the Mason Greenlee party in 1871, and others.[61] Historian Patricia Etter notes that "numerous forty-niners pocketed a copy of" Emory's notes, making stops on the Gila Trail to prospect along the way to California in search of Emory's gold.[62] The *Santa Fe Gazette*, on September 24, 1853, printed F. X. Aubry's journal about crossing northern Arizona, which included the statement that he traded old clothes for some Apache gold bullets. The gold bullet story went viral across the newspapers of the United States then Europe, even being read into the *Congressional Record* by Sen. William M. Gwin of California. Gold hunting expeditions lured by the myths of old trappers seeing streams "freighted with gold" or of gold mountains or Apache "gold bullets" stimulated the first placer gold rushes.

Gila City

During September 1858, gold seekers (who described themselves as argonauts) under the general guidance of Col. Jacob Snively, a former secretary to Sam Houston, were drifting back from disappointments in the California goldfields when they discovered placer gold along the Gila River twenty miles upstream from its junction with the Colorado River. The resulting rush established a completely different pattern for claiming mineral rights for Arizona's first gold rush.[63]

Evidence of any legal records in this area between 1858 and 1862 has not been found in the usual county or state records or repositories. But this was fortunately an era of letter writing, and some of the Gila City rules were mentioned in letters published in newspapers of the period.

California's Senator Gwin was quoted in the *New York Times* of January 5, 1859, on how miners at the Gila mines operate; he reported that the "Gila mines are placer diggings, wherein any man can operate if he can secure a claim, and is possessed of a pick, spade and tin pan." This apparently did not resolve disputes, however. A November 27, 1858, letter in the *Daily National Democrat* of Marysville, California, reported that a fight over tools sparked a miners' meeting, which selected five people to "form a code of laws for the rejection or approval of the people" and added that "the people of western Arizona if not in name, are in reality without any law." A January 15, 1859, letter from "Gila City, Arizona Gold Mines" published in the *Los Angeles Star* on January 29, 1859, reported that miners held another meeting and passed a law "prohibiting any one person from holding more than one claim," and that a claim must be worked every three days, or anyone may jump it. Claims were six feet wide and were in some twenty-two clusters that spread six miles along the sand hills above the river. By spring, a thousand miners worked the diggings, mining under "squatter sovereignty," as Senator Gwin described it. The last miners' meeting recorded in the press was to reverse an original rule that excluded Sonorans from the diggings, and was reprinted in the *Daily Missouri Republican*, January 28, 1859.

A miner known only as "Utah" wrote a series of letters about his party's mines and confirmed their taking a claim by "squatting" on a prospect they called "Stoddard's Diggings" after one of the members.[64] When claim owner Robert H. Birch had a townsite surveyed, christened Gila City, and lots were sold, the miners met again and elected J. G. Riddle chair. They also voted for a sheriff and judge from among their own to provide law in the diggings. Surveyor Ordeman, author of a letter in the *San Joaquin Republican* on December 15, 1858, noted the election of officers for the town, but petty violence and the theft of tools continued. When Sylvester Mowry visited Gila City during his campaign for delegate to the yet unrecognized Territory of Arizona, the miners met with Mowry and gave him a list of resolutions supporting a territorial government for the gold mines of the Gila.[65]

A judge named Ned McGowan was finally appointed to hold court in Gila City for the unrecognized Arizona Territory and arrived in June 1860, as noted in the *Daily National Democrat* of Marysville, California, June 14, 1860. But Gila City's population by that time had diminished to 157 enumerated in the precinct, according to the June 1860 census,[66] and the miners followed Jacob Snively in several expeditions prospecting up the Gila and Salt to the head of the Black River, up the Gila and Salt to the Verde Valley in 1859, and far up the Gila, past the San Francisco (Emory's Rio Prieto "freighted with gold") to a mountain tributary they called Bear Creek, where they claimed to have found a bonanza.[67]

With the beginning of the Civil War and increasing conflict with the Apache, Snively and others relocated back down to the lower Gila and began prospecting for placer and hard rock claims along it and the Colorado. By then another rush had begun at La Paz on the Colorado River.

Near La Paz on the Colorado River

La Paz, the largest of Arizona's 1860 gold rushes, began with Sonorans moving from the Gila City area up the Colorado River and discovering new but small goldfields at Laguna, Picacho, and elsewhere. The actual "rush" may have been sparked when Paulino (Pauline) Weaver showed gold dust to José María Redondo, a trader and miner at the Gila City diggings. Redondo, Felipe Amvisca, Antonio Contreras, and Jesús Contreras may have been the true discovers.[68] By February 1862, the word had spread but the interest was primarily within the Hispanic community. During April and May, the fever had hit Los Angeles as samples of gold reached the city and Juan Ferrar, a forty-eight-year-old Sonoran, had picked up a specimen reported to be the size of a "hen's egg." Then George Hooper, the trader at Fort Yuma, showed up in Los Angeles with $8,000 in gold dust.[69]

La Paz on the Colorado River was the hub. The placer goldfields stretched among the foothills south and east for twenty miles. One of the first to the diggings described the scene of three hundred miners, mostly Sonorans: "no claims, so-called, are allowed. A man is allowed space enough to use his pick or crowbar, or other implements, nothing more. Another may pitch in right alongside. These mines represented as giving more general satisfaction . . . [than] the opening of Mokelumne Hill, [California] in 1851."[70] One, "Californio" from Los Angeles, wrote home calling for more women to join them in the diggings, especially after she panned $300 in a few days.[71]

Herman Ehrenberg arrived in the district in 1862, and it appears that no particular laws either existed or had been memorialized when he arrived. During that year, Ehrenberg appears to have been responsible for the enactment of the first written regulations that followed the template of the gold-rush districts in California. These first regulations applied to lode (hard rock) claims within the La Paz and Castle Dome Mining Districts. The La Paz Mining District was formed October 6, 1862, to include the lands along the Colorado River beginning approximately ten miles downstream from the Bill Williams Fork. The Castle Dome Mining District was formed at a meeting held at La Paz on December 8, 1862, to organize the miners along the Gila River near the town of Gila City.

Snively, apparently in recognition of his longevity in the area, was elected chairman of the Castle Dome District, and Ehrenberg acted as secretary. The records of the La Paz District show that C. O. Cunningham acted as the first recorder.[72] An apparent reason for the delay in enacting these regulations was that the early activities of both the Castle Dome and La Paz Districts were strictly placer mining in which permanency and formality appeared to have been a secondary consideration. The miners simply did not have time to go through the formalities of staking a claim in their quest for the best spots to pan gold. The rule was not unlike that of the fisherman, which would allow one to work the place in which he is standing, but once a spot was vacated, the next comer was under no restrictions. In any case, the first regulations formally enacted in the La Paz District were for quartz claims and permitted a mining claim not exceeding two hundred feet in length "with all its dips, spurs and angles."[73] The width of the claim was not specified. In addition, the discoverer was entitled to an additional claim on "every lode he may discover." The process of location required the claimant or claimants to post a notice on the claim, and if a claim had been neither worked nor recorded within twenty days, it would be subject to forfeiture. The discoverer was also required to submit a specimen of ore from the claim to the recorder to be preserved for future reference by the recorder, who was required to be a resident of La Paz.[74] The regulations of the Castle Dome Mining District enacted two months later were patterned upon the prior experience within the La Paz Mining District but were much more detailed in laying out their requirements and permitted more land to be included within individual claims.[75] The Castle Dome District Regulations were also for vein deposits, and permitted a mining claim of one hundred yards in length along the vein, "including all the angles, spurrs, etc., belonging thereto," together with one hundred yards on each side of the vein so long as there were no conflicting rights. The discoverer's "bonus" was also recognized and allowed the discoverer an additional hundred yards on each separate vein so discovered.[76]

The first book of records of the La Paz Mining District covers a period from October 6, 1862, to March 4, 1863. During that time, a total of ninety-one separate claims were filed, which were virtually all by associations or groups of individuals filing a group of claims on the same lode. These groups were generally from four to six persons, and during this period approximately four hundred separate claims were located within the district. Most claims were not named, as was the Mexican and California tradition, but only numbered. The first claim within the district was in Spanish and was located the day before the formation of the district on October 5, 1862. This claim was named "La Costancia" by Marcilimo Encinas and was described as one hundred feet square. The claim was apparently part of a larger operation and was described as being in the "Murible Diggings at the American camp."[77]

The method of preparing location notices was by an allocation of the vein among the miners in a single notice giving the discoverer four hundred feet and each individual claiming two hundred feet, and listing the total length of the claim. For example, the fifty-ninth claim located read as follows:

> We the undersigned claim in this Lead 1200 feet commencing at this point four miles North West of Tatman Well with all its dips, angles, spurs for mining purposes. This lead was discovered the seventh day of January, 1863.
>
> | | | |
> |---|---|---|
> | Jacob Goldman | Dis. | 400 feet |
> | Tho. Freeman | | 200 |
> | W. D. St. John | | 200 |
> | F. G. Fitch | | 200 |
> | Jacob Rastel | | 200 |
> | | | 1200[78] |

Ehrenberg appears to have arrived in the district about this time and is named on a claim notice located on November 14, 1862, with Robert Mason listed as the discoverer.[79] The Colorado vein was staked thirteen miles south-southeast of La Paz east of the Colorado River on January 20, 1863, by Colonel Snively and seven associates. The claim was 1,800 feet in length and included "all the angles, dips and spurs belonging to the vein." Also 100 feet of ground was claimed on either side of the vein.[80] Ehrenberg was also listed as a locator of the Eugenie Quicksilver Mine, which was twenty-two miles southeasterly of the town of La Paz, on February 17, 1863.[81] It appears that the population of the La Paz District included a large Mexican population because approximately a quarter of the transactions within the La Paz records were written in Spanish. This fact is corroborated by Samuel Heintzelman's remark in his diary on December 23, 1858, that all the Mexicans employed at the Cerro Colorado Mine left for the gold mines on the Gila.[82]

Fig. 2. Herman Ehrenberg (1816–66). A mining engineer and cartographer, Ehrenberg was the organizer of the La Paz and Castle Dome Mining Districts. He was born in Prussia and traveled to New York as a teenager. In 1835 he enlisted with the New Orleans Greys on behalf of what would become the Texas Volunteers. He was one of the few survivors of the Goliad Massacre. Courtesy of the Wittliff Collection, University of Texas, San Marcos.

Considerable attention was devoted in the Castle Dome regulations to work requirements. Initially the problem of a lack of mining implements was noted, and a time for commencing operations on the claims was fixed at March 13, 1863. After this date the claims were required to be worked within thirty days and thereafter "in good faith for at least 4 days in each month." For companies holding various claims it was not necessary to work each one, but working on any one of the claims would be sufficient evidence of good faith for all the parties so claiming.

The maintenance of records for the district was under the authority of the recorder "as presented in a book kept for that purpose." Like the

La Paz District, the claimant was required to provide a specimen of ore to accompany the location notice. The purpose of the exercise was to provide evidence in case of a future dispute.

Elsewhere along the Colorado River to the north of Hardyville, a separate group of lode miners formed the Colorado Mining District sometime during the middle of 1862. Formal laws for the district were adopted at a meeting at the San Juan Company House in Eldorado Canyon on January 8, 1863.[83] Apparently Lewis had been previously elected as recorder on June 1, 1862, and was reelected to fill the unexpired term. Approximately thirty-five miners attended the meeting, as the minutes report that William Caley was elected president by a 20–8 vote over Dr. Bush, and Wilder defeated Poindexter for vice president in a 19–15 vote.

The district extended ten miles north and south along the Colorado River from Eldorado Canyon and included the twenty-mile strip on the west side of the river. Thus, the district did not include much of present Arizona, although at the time all of the area was included within New Mexico Territory and later Arizona Territory until much of which is now included in present Clark County was ceded to Nevada in 1866. The district's regulations were surprisingly sophisticated and placed substantial duties on the recorder. In addition to maintaining the books, the recorder was to verify that no conflict existed over claimed ground before recording a notice,[84] and it is clear that a physical inspection of claims was a part of the recording process. For example, the recorder was paid a filing fee of $2.50 for up to four claims and 50 cents for each additional claim for all locations within three miles of the recorder's office. If further travel was required, the recorder was paid $1 per mile.[85]

The laws stipulated that the location of lode deposits and quartz claims were not to be in excess of two hundred feet in length with previously located claims entitled to "fifty feet of all Spurs or Offshoots extended from their junction with said Lode and Fifty each side of surface ground for the purpose of operating said Lode."[86] In addition to lode claims, however, the district also recognized the legality of tunnel sites, and permitted any party to drive a tunnel that would give rights to all lodes that might be discovered in the tunnel that had not been previously discovered.[87]

One of the more innovative aspects of these laws was in the requirement for the performance of development work. The laws specified that recording would hold the claim for six months, but thereafter, one day's labor was required for each thirty-day period, although groups could consolidate work as part of a common effort. The laws also recognized the purchase of powder, fuse, and drilling and mining tools actually used in the operations as applicable toward this requirement, and the performance of one hundred dollars' worth of labor would hold a claim without further work until abandonment.[88]

The Arizona side of the river was not the sole interest of the adventurers. The Freeman District, formed on June 12, 1863, lay to the south of the Colorado District adjacent to the Colorado River on the California side.[89]

The Weaver Districts

The district's namesake, Paulino Weaver, a fur trapper, had been in Arizona off and on since the 1830s.[90]

In 1862 Weaver found gold ten miles north of present-day Ehrenberg, Arizona, along the Colorado River. This find was formalized in mining district regulations of the first of several mining districts to bear his name.[91] A public notice posted in La Paz on March 9, 1863, gave official notice of a miners' meeting in Olive City on March 20, 1863, to consider the formation of the district. On that date, some fifty-seven miners, after some debate, appointed a committee of James Reed, J. B. Chevelier, W. B. Marshall, H. M. Oliver, and Louie Robinson, who were given an hour to draft regulations and define the district's boundaries. These approved regulations formed a new Weaver District immediately to the north of the Castle Dome District and along the east bank of the Colorado River. These regulations allowed claims of two hundred feet in length and one hundred feet on each side of the vein at the surface. Upon the posting and monumenting of such a claim, the locator was allowed twenty days within which to commence work, and the failure to do so or make record of the same would result in the claim's being subject to forfeiture. Work requirements consisted of three days' labor every ninety days and twelve days within a six-month period. These regulations effectively ceded that portion of the older La Paz District that lay within the boundaries of the new district. This action emphasizes the underlying premise of mining district regulations that government exists only by the consent of those individuals willing to be governed. The regulations of the Weaver District were printed in the San Francisco *Morning Call* newspaper on April 9, 1863.

Weaver was then selected to guide the Abraham Harlow Peeples party, recent arrivals from California, from Yuma to the Prescott area as Weaver

Fig. 3. Paulino Weaver (1797–1867). A trapper and guide well known in Arizona and the West, Weaver filed the "Arivaypa" lode mining claim on May 29, 1862, fifteen miles north of "Old Fort Breckinridge" and two and a half miles southeast of the Cottonwood Spring, likely near the present Ray Mine. The notice was registered with the Arizona Territorial Secretary on August 24, 1864. Courtesy of the Sharlot Hall Museum, Prescott, Arizona.

had been quoted as having seen gold in the Verde Valley or San Francisco Peaks years before and was now ready to lead an expedition to the interior. At this same time, Weaver had organized a meeting of the tribes along the river to give free access to prospectors into their domain. An agreement was completed in April 1863, when Weaver left to organize his gold hunting expedition. Weaver party member M. W. Weber later stated that "if it had not been for this treaty the miners could not have gone into that country as they have done."[92] As the party moved upstream along the Bill Williams Fork and the Santa Maria River to Prescott, new mining districts sprang up. On June 25, 1863, a second Weaver Mining District was formed commencing at the mouth of the Hassayampa River, over to the southern boundary of the Walker District, then westerly to the Santa Maria River, then southerly ten miles beyond a crossing of Date Creek, then returning to the Hassayampa.[93] The claims of the new Weaver District were 150 feet and appeared to be intended to cover only placer claims. The claims were limited to 150 feet along the creeks or gulches and 75 feet on each side. Miners within the district were limited to one claim except that the original ten members of the group were allowed one additional claim. All claims had to be recorded within five days of the date of location, which gave the discoverer sixty days to marshal a work program, after which time the claim was to be worked one day in ten.

The framers of this new district wanted it to be as exclusive as possible and prohibited citizens of Mexico from holding or working claims within the district, except for one young man, Lorenzo Para, who was from Mexico and was one the original discoverers. Further, if any miners employed Mexican citizens, they were held responsible for their good behavior, the failure of which mandated the forfeiture of the claim and banishment from the district. The regulations further prohibited any transfer of claims for sixty days after the adoption of the laws and prohibited nonresidents from holding claims. The district was apparently also concerned with keeping out lawyers, and by its fourth article provided that "all disputes in reference to mining claims in this District to be settled by arbitration." The duties of the recorder followed the similar pattern requiring free access to the records and disclosure of ownership, which specified that no bills of sale or claims would be valid unless recorded within forty-eight hours after the date of sale.

In June, the Antelope or Weaver diggings were soon eclipsed by the discovery of gold on the hill between the two creeks. A new arrival with Jack Swilling and Bernardo Freyes (also Berardo Frerer) at an arroyo on top of what became known as Rich Hill found the richest gold placer or surface diggings of all the period's rushes.[94] Only ten claims had the rich pay ore, but within a two-month period Swilling had scraped up $20,000, a nugget of which he sent to General Carleton in Santa Fe with an appropriate return thank you note. Charles Genung reported that the claims were only 10 x 10 and the best ten paid grandly, including Weber's, one of the original party who had $40,000 within a few months. The famous Rich Hill, where members of the party were alleged to have scooped up gold nuggets with eating utensils, was also a part of this district, and special provision applicable for the Rich Hill Mining District included a provision whereby a placer claimant could claim as much ground as could be included within a circle with the radius being the length of a long-handled shovel. News of the find along with news of a second prospecting party's discoveries in the upper Hassayampa in the connected Sierra Prieta, Bradshaw, and Weaver Mountains brought the rush of the Hassayampers.[95]

The Walker Districts

A little more than a month before the creation of the laws of the second Weaver District, the first laws of the Pioneer or Walker Mining District were passed.[96] They were initiated at a miners' meeting held on the Oolkilsipava (not surprisingly renamed Hassayampa) River on May 10, 1863. The Walker District was to become the most active in the history of mining district regulations. This party, led by Joseph Reddeford Walker Jr., a contemporary and friend of Christopher "Kit" Carson, had begun its expedition in 1861 (at the time, Walker was sixty-three years of age) in California, crossed the upper Colorado River, and headed east to prospect the eastern slopes of the San Francisco Mountains where Walker had seen gold while trapping in the 1830s. During the early winter of 1862 the party was in Denver, Colorado. The invasion of Confederates into New Mexico in late 1861–62 caused the group to scatter, with Walker going to Colorado and a few joining Union troops to defend the territory. Agreeing to rendezvous in Santa Fe in fall 1862, the reorganized group with new members from Colorado and New Mexico received passport papers allowing travel through the territory—then under martial law—to "go into the country north of the Gila" to prospect. They left Santa Fe early December 1862 headed to the vicinity of the Rio Prieto, Emory's stream "freighted with gold." The group arrived in Pinos Altos by January and, joining with Union soldiers, spent a month prospecting the Prieto (or San Francisco River), where Walker said they found diggings—"up to 40 cents per pan"—but not as rich as he expected to find back in central Arizona. The Walker party left in April for a planned prospect down the Gila, up the Salt and then Verde Rivers. Along the way Jack Swilling convinced the party to divert to the uncharted Hassayampa River where he had seen gold a few years before—Weaver's just completed peace treaty negotiations would offer some easing of tensions with indigenous groups and allow them to explore the region. Walker wrote letters to John Moss on the upper Colorado and to Jacob Snively, and probably others, to explain their plans and invite them to join. The last of the major placer gold rushes was about to begin.

The group had no clear view of the legalities of claiming mineral ground and adopted whatever procedures seemed appropriate at the time. After passing through Tucson, the group proceeded to the Pima villages near Sacaton. There, on May 21, 1863, Albert C. Benedict, a member of the group, wrote to Kirby Benedict, the chief justice of the New Mexico Supreme Court, asking, among other things, for legal advice as to the proper way of protecting his rights as a discoverer of silver mines and requesting further information as to the boundary line of the newly created Territory of Arizona.[97] The advice given by the chief justice, if any, is not known, but Benedict did finally locate a claim on August 8, 1863, claiming 4,000 feet for himself and eighteen other individuals, including Judge Kirby Benedict and Judge Benedict's son.[98]

The laws of the Pioneer District were adopted at three miners' meetings held on May 10, June 10, and July 12, 1863. The lode claim provisions were formally carved out of the district with the formation of the Walker Quartz Mining District on November 24, 1863. Both the laws of the Pioneer and Walker Districts and their administration belie the traditional theory of the "plenty for all" view held by some writers in characterizing the California camps and are good examples of abuses that can be perpetrated by tribal self-government. The regulations of the district were, contrary to the California genesis, monopolistic. One writer has suggested that the reason for this was that the members of the Walker party were not miners, but were prospectors or adventurers and, in a sense, were speculators.[99] At the first meeting of the "Walker prospecting & mining company" on May 10, provisions were made for fifty-two placer claims that were one hundred yards long and fifty yards from each side. Each member of the twenty-six-man party was entitled to a single claim plus an additional claim by reason of being a member of the discovery party. The best areas were marked off and given numbers with the members of the party choosing claims by a lottery. All "outsiders" were prohibited from entering into the district until the original discoverers had a chance to select their claims, and a special provision was made that no Mexicans could locate or purchase any claims from June 1 to December 1, 1863. Newcomers were also discriminated against by way of a taxing structure whereby no fee was charged for recording the original claims, but all others paid $2.50 per claim. Disputes were to be resolved at a miners' meeting called for that purpose, in which case the disputants were required to pay $5.00 each with the prevailing party getting his money back and the losing party forfeiting the fee.[100]

The second meeting on June 10, 1863, both enlarged the district and allowed two additional days for the original claimants to select their claims.[101] At this second meeting, the first provisions were made concerning the marking of boundaries with a requirement that the corners of the claims be marked and numbered by blazing trees or using stakes. Ten days were allowed from the date of location for recording, and locators were prohibited from locating claims for third parties. Also, at the second meeting, the first provisions were made for lode claims that allowed claims of two hundred feet in length. There were no work requirements imposed by the law, and a special provision allowed the claimants to hold any claims for one year whether they were worked or not. Having controlled the Mexicans in the prior meeting, and apparently fearing that the area would be attracting Chinese laborers from California, the regulations enacted at the second meeting prohibited "Chinamen" from taking up claims within the district. Problems still existed regarding Mexican labor, however, and a further provision was made that if anyone employed Mexicans, the names of the individuals had to be placed of record with the recorder, and the employer was to notify the recorder when such individual was discharged. A one-dollar filing fee was charged for each name.

By July 12, the date of the third meeting, pressure was being brought to reconsider the Mexican issue, and by the first resolution it was stated that the laws regarding Mexicans remained the same. However, in an apparent concession, a committee of three was appointed "who shall decide who are and who are not Mexicans."[102] At this third meeting, the party was still not satisfied that they had taken up the best ground, and an additional ten days was allowed for the original party members to locate claims anywhere in the district. A further provision was made prohibiting more than one claim on each stream. A committee of three was appointed to measure claims, and ten days were allowed within which to record notices of location upon payment of $1.50 fee. "Asiatics and Senoranians" continued to be excluded.

A major reorganization of the district took place at the meeting of the miners on November 24, 1863. At this meeting the president and secretary tendered their resignations, and new officers were elected. A committee of five was appointed to draft laws governing quartz mines, and a committee of three was appointed to draft new laws to govern placer mines. The five appointed to recommend the lode laws consisted of Van C. Smith, Soloman Shoup, Cal Dobbins, Major McKinney (probably Rodney McKinnon), and J. M. Sanford. This group had given the matter advanced thought and immediately submitted a report that began with a defense of criticism that was apparently being directed toward the original members of the Walker party. This statement read that they would, under all circumstances, defend and protect, aid and assist all traders and persons whether citizens or not for the persecution of right and legitimate business within the jurisdiction of the laws of the Walker Mining District. The report went on to "denounce the originators of the many falsehoods, circulated by the fainthearted many who have returned to their shin warming firesides as a set unworthy the name of Pioneers . . ."[103]

With that off their chests, the report proposed laws following the traditional California models fixing the length of lode claims at two hundred feet along the lode and fifty feet on each side and including all dips, spurs, and angles following the ledge. The discoverer was entitled to one claim as a "discovery claim" plus one additional claim. The locators were required to erect monuments either by stakes, ditches, furrows, stones, or trees at each end of the claim; these notices were required to designate the date of location, the amount claimed, the names of the parties, and the direction following the ledge. Sixty days were permitted for recording from the date of posting notice, and the claimant was required to furnish the recorder with a true copy of the notice and a specimen from the mine. By the thirteenth article, provisions were finally made requiring performance of some work to maintain claims. This article specified that three days' labor be performed on every claim every three months for one year. Some of the restrictive provisions remained in prohibitions against anyone but "white persons" from holding claims within the district, and the provision also prohibited nonresidents from holding claims. Further, no person except the discoverer could hold more than one claim on each ledge except by purchase. All disputes within the district were required to be settled by arbitration; each party to a dispute was to select one arbitrator, and the two arbitrators selected a third. Persons acting as arbitrators were entitled to a fee of five dollars per day, a fee that had to be first deposited by the parties calling the arbitration with the president of the district. The losing party would pay all expenses, and the money deposited by the opposite party upon demand would be returned to them.

The three members appointed at the November 24 meeting to draft placer laws either never presented a report or their report was lost.[104] It appears, however, that the existing placer regulations of the Pioneer District remained in force, and thereafter the government of the district was provided through parallel rules for the Pioneer Placer District and the Walker Quartz District.

With the more "conventional" set of regulations represented by the November 24 meeting, legislative efforts of the Pioneer and Walker Districts peaked. One further attempt was made to discriminate at a meeting held on May 22, 1864,[105] wherein article 16 of the laws of the Pioneer District was amended to prohibit any person "within the service of the United States to locate a claim within this District." Such a provision would have included all of the territorial officers who were appointed by the president of the United States, and would have included some officers who were quite interested in mining, not the least of which were Governor Goodwin and Territorial Secretary McCormick, who had already been named as owners of a mining claim within the Weaver District. This also affected General Carleton, who was charged with the protection of the area. Whether this requirement was a mistake or an effort to correct some perceived evil is not clear. What is clear is that the prohibition was repealed a mere nine days later at a meeting on May 29, 1864. The restriction on Mexicans, however, was extended for an additional six months at a meeting on May 22, 1864.[106]

The area around Prescott continued to be very active during the remainder of 1863 with three new districts—the Yavapai, Hassayampa, and the Quartz Mountain Districts—formed on September 10, December 6, and December 27. All followed the pattern that had been set by the regulations of the Pioneer and Walker Districts.

Upon its formation, the Yavapai District elected A. B. Smith as president and George W. Leihy as secretary. The organizers first appointed a committee to draft the laws and boundaries of the district and gave them one hour to make a report.[107] The result followed the traditional requirements with a limitation of one claim per locator on each vein with the exception of the discoverer who was entitled to one additional claim. The laws addressed only lode claims and permitted claims of a length of three hundred feet along the strike of the vein and one hundred feet on each side following the dip. A notice of location was required to be posted in a conspicuous place and recite the direction of the claim. Each locator had thirty days within which to record his notice with the recorder, and in the event of failure to do so, the claim "shall be deemed abandoned and subject to relocation by any other party."[108] A similar provision also applied concerning a work requirement that required the claims to be worked within sixty days or face the same penalty. A work requirement was imposed requiring three days' work in every ninety or, alternatively, twelve days in six months would hold the claim for one year.[109]

The laws of the Hassayampa District,[110] with Robert W. Groom acting as the recorder, provided for only lode claims that were 300 feet in length, together with 150 feet on each side "next to the lode, with all minerals therein contained."[111] At a subsequent meeting of the district held on May 10, 1864, the time for performance of the first work was extended until October 1, 1864, due to the fact that King S. Woolsey was organizing a punitive expedition against the Apaches, and the miners of the district were being called upon to participate. At this meeting Van C. Smith, who was also president of the Walker District, was elected president.

The Quartz Mountain District was organized at a meeting held at George Lount's cabin on Granite Creek with John West chosen as president and C. M. Dorman as secretary.[112] The dimensions of lode claims were fixed 300 feet in length and 150 feet on either side of the vein, including all minerals contained therein. The locator was required to state in the posted notice the number of feet claimed in each direction from the point of posting, and to record this notice within sixty days. If this recording was not done, the claim "shall be deemed abandoned & subject to relocation."[113] The requirements were also traditional in permitting only one claim per locator on each vein except for the discoverer's two claims. A nontraditional part of the Quartz Mountain District laws concerned work requirements and established the only laws that purported to vest the locators with perpetual title upon the performance of a specified amount of work. Under these laws, each locator was given until May 1, 1864, as a grace period, but if the claims was worked for six days before that date, a "perpetual title" would result. This perpetual title was not entirely "perpetual," and the claimant was required to "renew" his title after two years—or it would be lost. In all other cases, work requirements consisted of three days' labor in every ninety, but if an individual kept up this requirement and performed twelve days' work in a year, he would also obtain a perpetual title. In all cases, the recorder was required to inspect the work performed, and if the work was satisfactory, a certificate of the performance of work was issued for which duty the recorder was paid one dollar for each claim.[114] Finally, a second nontraditional aspect of the Quartz Mountain District's laws was an authorization permitting all persons holding claims within the district to vote, including nonresidents who were permitted to vote by proxy.

In the case of each of these three new districts, the district was required to be run by meetings of the miners that could be held on ten days' written notice posted in three conspicuous places within the district.

The Silver of Southeastern Arizona

In the southeastern part of the what would become Arizona Territory, the absence of military protection during the early days of the Civil War and the danger from Indian attack stifled the efforts of miners during the period after the shutdown of the Cerro Colorado and Santa Rita Mines in July 1861 until Arizona's new territorial government was in place in 1864.[115] With the establishment of official records in the First Judicial District in Tucson, the attempts by miners to legalize their prior efforts began to surface, and the first documents filed for record were petitions for official sanction of mineral claims. The first document of record is probably the most instructive of both the history of mineral development in the area since the late 1850s and the legal thinking of the prospectors:

> Territory of Arizona
> First Judicial District
>
> To the Hon Charles Turnbull Hayden, Judge of Probate for the District and Territory aforesaid,
>
> Your petitioners W. Claude Jones a resident of the District and Territory aforesaid, Col. George W. Bowie of the 5th Infty. California Volunteers Commanding the military District of Arizona, and Captain Charles A. Smith of the same regiment of Volunteers, respectively represent that they claim by right of discovery, two veins of argentiferous Galena, and one vein of the sulfide of Silver; one vein of carbonate of Copper and two ledges of gold-bearing Quartz on the sides of the ravine leading to the Dragoon Spring (the principal vein being on the right bank of the road leading from the old overland mail station to said spring) and on the western summit of the mountain commonly known as the Dragoon Spring Mountain, which they have called the "Monteverde Mines" and which they are desirous of working and developing to the extent of their means and ability by machinery and otherwise; said mines were discovered by W. Claude Jones in the Month of June 1858 and reopened in 1861 and 1864 but have not been continuously working owing to the hostility of the Apache Indians. Wherefor your petitioners being now desirous of organizing a company and providing

> systematically to work in developing said mines according to the "ordinanzas de Mineria" in force in this Territory, they respectfully pray your honor to denounce to their use and benefit, for the purpose of operating successfully in said mines, the corresponding two hundred varas for each of said veins, with their "pertinencias" which appertain to the right of discovery so as to extend around and include each vein of said "Monte Verde" Mines so discovered; and each of said Quartz ledges, together with the benefit and use of all the ores, minerals, earth, wood, water and grass, and particularly the spring of water known as the Dragoon Springs in the following limits to wit; . . .[116]

Jones had previous experience in the location of mining claims in the Organ Mountains of New Mexico in 1858,[117] and led the surge of interest in the Dragoons. His claims continued to follow the same pattern as his Monte Verde claim when, on July 9, 1864, again citing the authority of the *Ordenanzas de Mineria*, he located a gold mine with Manuel Gandasa and Henry McC. Ward "near Heuvari"; they alleged that it had been abandoned since 1849.[118] During August, two more claims were located near "Apache Pass," south of Fort Bowie, bearing the generic name of "Miner's Claim" and included ten and seventeen locators, respectively, who claimed 5,400 and 4,300 feet along the strike of the vein and 100 feet on either side. Individual portions were not specifically indicated but were to be 200 feet along the vein for each miner; no explanation is apparent for the excess unallotted length.[119] Claims of 3,300 feet in length were common and twenty-one of such location notices appear of record followed by a report of survey for the claims recorded. These records were apparently prepared by H. S. Washburn who had started a "private registry opened for the convenience of mines in the middle and western portions of Arizona."[120]

These 1864 records frequently also referred to the location notices as "record of demarkation and possession papers," and included mining claims near known mines in Arivaca by M. A. Davidson, W. S. Higgins, William Hartley, Elihu Baker, and William Sim during March 1864, and claims near the Babaquivari Mountains by W. Claude Jones, George W. Bowie, and Charles A. Smith.[121]

The belief in the legal efficacy of the *Ordenanzas de Mineria* was also superimposed upon the concept of the enactment of the "miner's rules" in southeastern Arizona, when on April 23, 1864, the Cerro Colorado District was formed, with William Sim elected chairman and John A. Mahon as secretary. The laws that were enacted at this meeting can be characterized as discouraging anyone who was not entirely serious about engaging in mining.[122] The boundaries of the district were established as a twenty-four-mile square around the Cerro Colorado Mine, and the laws were most concerned with defining the duties of the recorder. Office hours were established as between noon and 1:00 p.m. on each day of the week except Sunday, and the principal duty was to maintain the book of records. For this effort, the recorder was entitled to collect a recording fee of one dollar for each document plus fifty cents for each additional name on an instrument. These records were open for inspection for anyone "legally entitled to examine same" and upon the payment of a fee of one dollar. One might imagine that the recorder must have done very well for one hour's work per day.

The claims within the district were required to be 300 feet per individual along the vein and 300 feet on either side of the center of the vein. The laws of the district, however, permitted group claims where the claim was located by several individuals of 300 feet per individual up to a total claim length of 5,000 feet. The work requirements of the district were clearly the most rigorous of any district in Arizona Territory. Initially, the locator was permitted 180 days to secure tools and then 30 additional days within which to perform the requirements "of the laws of the United States, and the laws and regulations of this district." Here the framers of the laws were anticipating some federal legislation as there were no applicable laws of the United States at the time. The laws then went on to require the sinking of a thirty-foot shaft or tunnel in addition to six days' work for each claimant. This work would hold the claim for nine months, at which time work could be suspended. However, that if work was not commenced at the end of the nine-month period, the claim was "considered abandoned and open for location anew."[123]

The laws of the district could be amended upon a two-thirds vote of miners attending a meeting of the miners called for such a purpose. The call for the meeting required a ten-day notice signed by five members and was required to be posted in "Arevaca and Soppho." Finally, the importance of the public record was stressed, with a requirement to place claims of record within ten days after location and that "no claim shall be considered perfected until recorded."[124] These provisions are significant because of a link to the Spanish heritage of the Cerro Colorado District as exemplified by the requirements for a thirty-foot shaft and withholding perfection of the claim until recording.[125]

The Maturing of the District Rule

The year 1864 would see the authority of the autonomous mining districts challenged by the birth of a central territorial government in the heart of the lode mining districts. The new Territory of Arizona was created on February 24, 1863, and the territorial government was established on December 29, 1863. At that time, there were four principal mining centers. The most active at the time was in and around the area of what was to become Prescott and soon became the territorial capitol. The second center was along the placers of the Colorado River in Yuma and Mohave Counties at La Paz, Gila City, and Hardyville; third, along the Hassayampa River near Wickenburg; and finally, in southern Arizona.[126]

The Prescott area saw the formation of both the Walnut Grove Quartz and the Wickenburg Districts on May 21, 1864; Turkey Creek District on August 10, 1864; Bradshaw District on September 14, 1864; and Woolsey Quartz District on September 30, 1864.[127]

The Walnut Grove Quartz District was governed by a single officer, the recorder and a deputy (F. M. Larkin, the deputy recorder, signed the minutes of the miners' meeting) and was located south of the Hassayampa District and east of the Weaver District. In most respects, this district was similar to those that went before it. Claims were 300 feet "running with the main lode," and 150 feet on each side.[128] The locator had sixty days within which to record and complete the location process by posting a notice in a conspicuous place indicating the number of feet claimed each way from the location monument. The starting point was always considered as the discovery claim. All locators were, however, required to be a "sitizen" of the United States.[129] No work requirements were imposed, and if the claim was properly located, it would "be deemed sufficient to hold such ground and give . . . perpetual title thereto."[130]

The procedures for making changes in the laws of the Walnut Grove District were a little different and required that any changes had to be published "for three months in any paper published in this Territory" and posted in three conspicuous places at least ten days prior to the meeting and signed by ten quartz miners.[131] All owners could vote, and nonresidents could vote by proxy, provided that they owned at least one claim.[132] There is no indication, however, that votes were weighted by the number of claims owned.

Probably the most important mining activity in the early life of Arizona Territory was the Vulture Mine. This gold deposit was first discovered about the first of November 1863 by Henry Wickenburg (born Johannes Henricus Wickenburg), a Prussian prospector who was attempting to catch Paulino Weaver and the Peeples party and who had missed out in the discovery of Rich Hill. Wickenburg and several partners, after staking the claim, went on to the Prescott area. A slightly different group returned seven months later, restaked a new claim on May 21, 1864, and formed the Wickenburg District.[133] Wickenburg was elected president, and James A. Moore was elected secretary and became the recorder.

The claims within the Wickenburg District were to be three hundred feet "running with the dips and angles of the lode."[134] Obviously, Moore did not copy the laws upon which he was relying too carefully as it is difficult to conceive of a claim running with the downward dip of the vein or taking off at the various angles that could not have been known from the surface. One hundred fifty feet was also permitted on each side of the vein. The provisions concerning posting of the notice, contents, and the concept of one additional claim for the discoverer followed the standard format. Locators were permitted thirty days to record and "after expiration of 30 days, such notice be not found recorded, or filed with the recorder for record, the ground so claimed shall be subject to re-location."[135] The locator was required to do five days' labor within ninety days after the date of location, which then held the claim for one year.[136]

It was apparently possible to locate claims on behalf of others so long as the individual was acting as an agent for a resident of the territory, and all owners of claims had a vote within the district. A special provision was also made for a "water privilege," which if taken up by notice and recorded was good for a term of one year.[137]

In the formation of the Turkey Creek District, Charles Taylor was chosen as its first president, and W. C. Collier acted as secretary of the initial meeting and continued in the office as recorder. The initial regulations indicate that a previous code had been in existence, as most of the provisions were repealed. However, these laws have not been found.

The Turkey Creek laws followed the pattern of Walnut Grove as the recorder was the only officer of the district and was elected for one year with basic duties to maintain the books and keep them open for inspection. All inspections of records had to be in the presence of the recorder or his deputy, and recording fees were usually one dollar. Mining claims followed the earlier pattern established of 300 feet long and 150 feet on each side of the lode, but a pattern began to emerge that gave a little more definition to the rights of the miners within the limits of the claim. In describing the limits of the claim, the laws stated that the length of the claim would follow "all its dips, angles and variations, and all the minerals contained with the same one hundred and fifty feet, (including all spurs or parts of spurs of the same lode) shall be considered as claimed by and belonging to the locator or locators of said main lode, and part of parcel of the same lode."[138]

The locators were required to record a notice of location within sixty days that would be endorsed by the recorder on the back. Claims that were not timely recorded were "considered virtually abandoned and subject to re-location." Locators were limited to citizens of Arizona, but "miners or heirs" could hold claims by purchase. Any claim holder could vote in district elections, but proxy voting was prohibited.[139] The rights of the miners to claims were finalized by a "certificate of ownership" that was issued upon demand twenty days after recording but not later than six months, during which time all disputes must have been settled, and the claimant must have actually been on the ground.[140] Any disputes concerning the ground were required to be settled by a meeting of the miners.[141] The laws of the district did recognize the possibility that an individual might be away from the district for six months and therefore not able to obtain the requisite certificate of ownership; thus, provisions were made for exigencies beyond the control of the locator as well as detention in the service or armies of the United States. A good measure of the population of the Turkey Creek District was found in a provision that twenty voters were required in order to constitute a quorum.[142]

The Bradshaw District included the land in and around Montezuma City. At the organizational meeting, James A. Moore, now becoming a familiar face in the mining district organizations, acted as the chairman, and Dr. George M. Willing was appointed secretary. At this meeting, Max Solomon was elected president, and Dr. Willing was elected recorder; they constituted the only officers of the district.[143]

The laws of the Bradshaw District give the first indication of an awareness of miners and the possibility of the adoption of federal laws governing mineral rights. Thus, the laws required the president to faithfully cause the rules to be enforced "until the government of the United States shall, by act of Congress or otherwise, establish laws regulating the mines." The president was also required to act as ex officio judge of the miner's court before whom "all causes relating to the mines shall be tried."[144]

The laws of the district also showed the widest variation in different types of mining claims. The provisions concerning recording of notices included fees for (1) a surface or wash digging, (2) a placer claim, (3) a ledge or lode claim, (4) a millsite or lot claim, and finally (5) other claims not enumerated. The millsites or lots were obviously considered evidence of the existence of higher economic interest, as the fee for recording of such a claim was five dollars as opposed to one dollar for every other type of claim.[145]

In defining the size of the various claims and limitations on ownership, some further classifications of mining rights were indicated. For example, 300-foot square claims were recognized for discovery claims of all kinds, a claim by location, a placer or surface claim, a hill claim, and a lode, vein, or ledge of gold, silver, copper, tin, or other metal of value. A lot for building purposes was limited to 50 feet by 150 feet, but claims for water were three hundred yards square. The millsites for each mill or reduction works were one hundred yards square, provided that no claim could be taken up for agricultural, horticultural, or ranging purposes to the exclusion of mining operations. Of these various claims, each individual was allowed one claim by location on the vein (unless he was the discoverer, which allowed him an additional discovery claim), and one of the other types of claims had been described previously. The laws also recognized the possibility of conflict between ranchers and miners and provided for the superiority of mining claims so long as "all improvements of value on such lands or parcels of land shall not be interrupted or damaged, but where a valuable mine is known to exist and any such improvements be placed thereon the damages shall be assessed by three disinterested persons and the claimant of such mine shall pay to the owner of such improvements the assessed value of such damages."[146]

Specific provisions were made permitting two or more persons to form themselves into a company for both efficient working of the mine and as protection. In such an event, the company was permitted to hold up to as many claims as the individuals were allowed to hold in their individual names. Any miner was also allowed to acquire and retain additional claims by transfer, devise, or purchase. Any transfers, however, had to be by deed acknowledged before the recorder of the district or other qualified officer. When the party transferring the interest resided outside of the district, the transfer was required to be acknowledged before a judge of any court of record in the United States or the territories.[147]

The government of the district required that any alteration or repeal of the laws of the district could be initiated by eight members of the district filing a petition with the president, who would then call a special meeting by posting six notices in the most public places within the district ninety days before the day called for the meeting. At the meeting, only the subject matter contained in the petition could come up for a vote. However, further changes could be made at the meeting, provided that two-thirds of the miners within the district consented and were present at the meeting, and provided, further, that such change or repeal did not affect any claim or rights owned by persons under the existing laws prior to the alteration.[148] Every miner was entitled to a "seat" at any meeting provided he was a citizen of the United States and qualified under the laws of the district (which presumably meant that he owned an interest in the claim and also suggests that voting was by seats). A registry of members of the district was kept by the recorder.

Although no specific work requirements were imposed by the regulations, it appears that the miners considered some work necessary by their passage of an amendment at the organizational meeting that made note of the "danger apprehended by depredations of Indians and the safety of the mining community" and suspended any requirement to work in a mine for a space of one year between September 14, 1864, and September 14, 1865.

The last district that was formed before the adjournment of the first session of the Arizona Territorial Legislature was the Woolsey Quartz District, which had its initial meeting on September 30, 1864. At this meeting, John Roberts acted as chairman, and James O. Robertson was designated as secretary. In accordance with what then appears to have been tradition, Robertson was elected the recorder of the district. The Woolsey District included the town of Big Bug and vicinity—and was named after King S. Woolsey, a local rancher who was instrumental in organizing a punitive expedition against the Apache who had been engaging in "depredations." Perhaps the miners felt that Woolsey would provide more protection within a mining district named in his honor.

The laws of the district were parallel to those of the Turkey Creek District in both the location process and the idea that a certificate of title would be issued. The size of the claims were smaller, however, being only two hundred feet "on a level, including all dips, spurs, angles and variations, with 50 feet of ground on each side of the ledge or lode, and following all the dips, spurs and angles, with all the mineral contained therein."[149] Specific provisions were made, however, concerning the working of the claims. For all claims located after October 1, 1864, the laws specified that "each company locating ground together will be required to sink a shaft to a depth of ten feet, and of sufficient width to describe the size of the ledge." The digging of the shaft was required to be completed within sixty days of the date of location.[150] It is not clear whether the requirement for digging the shaft applied to individuals, but it is clear that some sort of work was required because the laws elsewhere provide for the issuance of a certificate of title by the recorder upon completion of work, which would allow the claimants to suspend work for six months, and if work was not commenced after six months, the "claim shall be declared abandoned."[151]

It is apparent that most of the regulations had many common threads related to recordation, the maintenance of records, the size of claims and limitations on ownership, and a concept of government by majority vote of individuals actually residing within the district. The rules seem to differ most widely in whether work would be required and the time for performance when required. One aspect of the district regulations enacted up through 1864—conspicuous by its absence—is any clear indication of any right to pursue the downward dip of the vein outside of the sidelines of the claim as marked on the surface. In virtually all of the district laws where even an oblique reference was made to extralateral rights, the only statement is that in addition to the main vein, the claim would include right to all "dips, spurs and angles," and all the minerals within the lateral extent of the vein. Only the laws of the Turkey Creek District suggested that the locator could follow all of the dips, angles, and variations of the vein, but even there it does not specify whether this right could be pursued outside the surface boundaries of the claim.

The First Litigation

Henry Wickenburg's discovery had the distinction of giving rise to the first significant litigation over mining claims in Arizona Territory.[152] When Wickenburg located his Vulture claim, absent any formal mining district covering the area, and also absent any territorial law, the locators posted the following notice on the claim:

> Notice—The undersigned claim this lode or ledge, the Vulture, with all spurs, dips, and angles of minerals contained therein. And according to laws of this Territory, Arizona. This November 24, 1863.[153]

The notice was signed by six individuals including Wickenburg, and his partner, E. A. Van Bibber. Other names on the notice included T. B. Green, W. Smith, N. K. Estill, and A. Fisher.

Green, who was also known as Theodore Green Rusk, sold his interest to William R. Murray and William Roberts, who returned to the claim on May 9, 1864, some two weeks prior to the return of Wickenburg and his associates. When the latter group arrived, they refused to recognize the ownership of Murray and Roberts, relocated the claim, and created a new mining district on May 21, 1864. Murray and Roberts thereupon sued based on the original location in what was probably the first litigation involving mining claims within the Territory of Arizona. Their case, styled *William R. Murray and William Roberts vs. Henry Wickenburg, J. A. Young, J. K. Simmons, James A. Moore and Valentine Griegerick*, was heard by the United States District Court for the Third Judicial District of the Territory and alleged that the original rights had been initiated pursuant to the mining laws of Spain and Mexico existing and in force in the territory at the time of the November 1863 location and that the plaintiffs were entitled to an injunction to prevent the defendants from interfering with the possession of the plaintiffs. The allegation that the laws of Spain and Mexico had any efficacy in Arizona Territory was a mistake. The case was heard by Judge Joseph Pratt Allyn, who was sitting in the stead of Judge William F. Turner, whose duties would normally have entailed cases arising out of the third judicial district.

Apparently, the only compilation of the Spanish mining law at the time was found in a decision of the United States Supreme Court in the *New Almaden* case, which Judge Allyn reviewed in the course of his decision.[154] He noted a requirement for a thirty-foot (ten varas) shaft and a requirement to present a claim to the mining deputation within the district. Judge Allyn concluded:

> The absence of "legal organizations and proper tribunals in the Territory" can scarcely be presumed to have prevented the attempt at least, to sink the shaft or well, specifically required by the mining laws alleged to exist and be in force in this Territory. There is no allegation

> that such work was ever attempted, or that there existed any obstacle to prevent its being done.
>
> The absence of "legal organizations and proper tribunals" would excuse a failure to register the mine, because the law does not require impossibilities, but the moment the impossibility is removed, it becomes then more clearly the duty of those desiring to acquire rights under the law, to obey the law . . .[155]

The judge went on to note that the civil government had been established on December 30, 1863, by a proclamation on April 9, 1864, a proclamation that formed judicial districts and assigned judges. The district court was held in Tucson in May and in June in La Paz, and thus the claimants had the ability to properly register the claim. Judge Allyn noted that "registry is the indispensable prerequisite of any title to a mine under the Spanish law"; he cited the commentaries of Gamboa to support this principle.

The injunction was therefore denied, and Judge Allyn held that no rights had been established under the initial location. Judge Allyn's reasoning suggested, however, that prior to the establishment of any formal mining districts and the recognition of any legal validity of such districts, the laws of Spain and Mexico regarding denunciation of mining claims were at least arguably appropriate as the basis for the assertion of rights to mineral lands. This point was unnecessary to the decision, however, because the complaint, although it asserted rights under the Spanish laws, did not allege that Green had complied with these laws. Judge Allyn summarized his holding as follows:

> It is not necessary for the purpose of determining the question raised by this prayer for a writ of injunction, to pass upon the question of whether the mining laws of Spain and Mexico, are in force in this Territory. It is sufficient that the complaint itself does not show that those laws have been sufficiently complied with, to vest in the complainants such a right of property in the Vulture lead, as to call for the extraordinary exercise of the chancery powers of this Court in granting a writ of injunction. The prayer for the writ of injunction is therefore denied.[156]

The entire text of Judge Allyn's decision was published in the *Arizona Weekly Miner* on October 26, 1864. The legal import of the obiter dicta regarding compliance with Spanish and Mexican law was not lost on

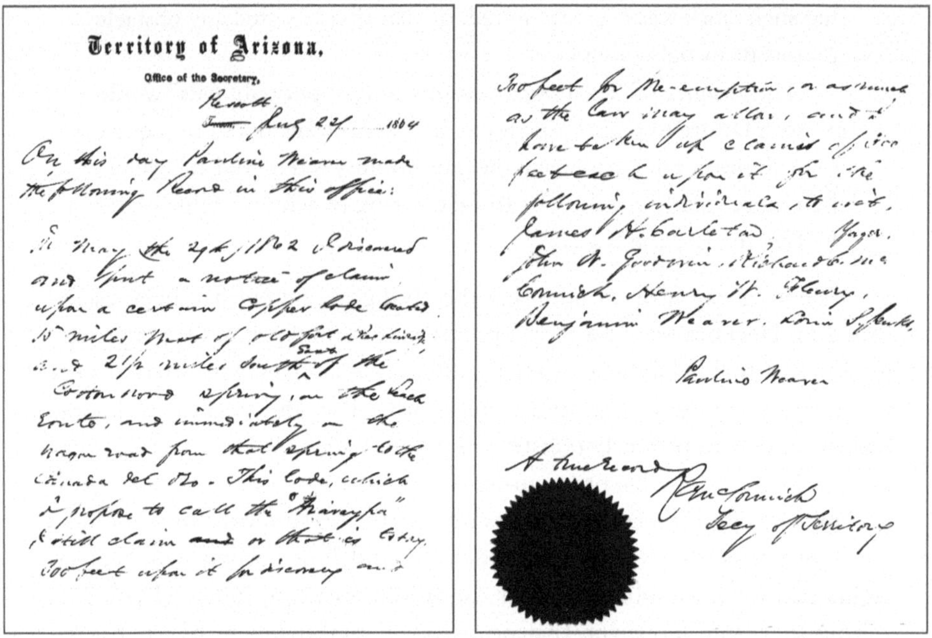
Territory of Arizona,
Office of the Secretary,

Fig. 4. Weaver's location notice from office of Richard McCormick. Records of the Territorial Secretary. A transcription of the text follows:

Prescott, Aug. 24, 1864

On this day Paulino Weaver made the following Record in this office:

On May the 29th 1862 I discovered and put a notice of claim upon a certain Copper Lode located 15 miles north of old Fort Breckenridge and 2½ miles south (east interlineated) of the Cottonwood spring on the Leach Route, and immediately on the wagon road from that spring to the Canada del Oro. This lode, which I propose to call the "Aravaypa" I will claim or that is to say 300 feet upon it for discovery and 300 feet for pre-emption or for as much as the law may allow and I have taken upon it for the following individuals, to wit: James H. Carleton, [space left for insertion of first name] Yeager, John W. Goodwin, Richard McCormick, Henry W. Fleury, Benjamin Weaver, Lorin S. Jenks.

Paulino Weaver

A true record
R McCormick
Secy of Territory

the miners, and Paulino Weaver appears to have immediately sought out Richard McCormick, the new territorial secretary, and swore out a notice of location on his 1862 claim that would have crudely complied with the *Ordenanzas de Mineria*.[157] Judge Allyn's decision underscored the belief of the miners that law did exist—and the best bet was that the law required compliance with some procedure that would demonstrate the good faith of a decision to work a mineral discovery, as opposed to simply a discovery. The general philosophy appears to be that the laws of Mexico applied, but even if they did not, there was a natural law of preemption that would rise to a defensible right when official laws were enacted.

Thus, in 1864, Arizona's miners clung to a tenuous legal thread of group self-protection enforced by the bylaws of some twenty formally organized mining districts. The final legal authority was still open to question, however, but the miners were betting that either the Mexican *Ordenanzas de Mineria* would be found to be a legitimate basis of a legal claim or that the United States government would recognize some sort of preemption or "squatter's" right under a new federal law. The legislature of the Territory of Arizona would get into the act first.

CHAPTER 3

Challenges to the District Rule

The General Mining Laws of the First Arizona Legislature

The two-year period between 1864 and 1866 brought a sea change to the autonomy that the miners enjoyed—a development that initially took the form of a law of general application enacted by the first legislative session of the newly created Territory of Arizona, and secondly was the long-awaited action of the federal Congress. Both events were closely followed by the miners in Arizona. The first effective legal dominion of the United States came to Arizona when a small contingent of federally appointed officials crossed the border into the newly formed Arizona Territory on December 29, 1863.[1] Arizona Territory was created as the western half of New Mexico Territory by an act signed by Pres. Abraham Lincoln on February 24, 1863,[2] and, since the laws of New Mexico Territory would have been in force within the area since its effective separation from the Republic of Mexico in 1846, the official body logically attempted to obtain copies of these laws on their way through Santa Fe. Only one complete set could be found and was kept by Richard C. McCormick, the territorial secretary. McCormick had ambitions beyond the position of secretary and established his political base immediately upon his arrival in the new territory through a newspaper he named the *Arizona Miner* in an obvious recognition of what would appeal to the existing population.

One of the first orders of business of this new government was to organize a legislature and create a code of laws for the new territory. As for the laws, the New Mexico model was rejected out of hand; McCormick was probably in the best position to know when he, as the editor of the *Miner*, observed in the July 20, 1864, edition that "the New Mexican statutes are not only crude and incongruous in the extreme, but the order or disorder in which they are printed leaves much to the conjecture of the reader. The books are irregular in form and plan, miserably indexed, and, for the most part, shabbily printed. While the laws reflect no particular credit upon their authors, the manner of their publication is highly discreditable."[3] He went on to express the hope that a new code "carefully prepared" by Judge William T. Howell, an associate justice of the Territorial Supreme Court, would quickly supplant the New Mexico statutes.[4]

Judge Howell had been given the task of preparing a code of laws by the legislature on Friday, October 1, 1864, and produced a draft eventually published in a volume of 680 pages, on Monday. In spite of any inference to the contrary from this sudden appearance, Judge Howell had obviously given the matter some prior thought. In fact, Judge Howell left Prescott for Tucson eight months earlier, on February 3, 1864, for his assignment as judge of the First Judicial District, and during the spring and summer drafted what was to become the Howell Code with the assistance of Coles Bashford, the ex-governor of Wisconsin, who was then an attorney practicing in Tucson.[5]

Fig. 5. Coles Bashford (1816–78). One of the principal authors of the Howell Code, the first legal code of Arizona Territory, Bashford was a former governor of Wisconsin (1856–58) who was practicing law in Tucson during an extended visit with William T. Howell during the spring and summer of 1864. Courtesy of the Arizona Historical Society, #23779.

At the beginning of the first legislative session, Gov. John N. Goodwin had some specific comments related to the mines and stressed the need for general territorial regulation of "quartz mining," meaning the mining of lode deposits. His message emphasized that the existing confusion between the various mining districts needed to be rectified in order to give the capitalists "clear and uniform titles." Placer mining, on the other hand could continue to be used by the various districts.[6] Judge Howell's proposals relating to mines and acquisition of mineral rights had obviously been communicated to the governor, and the governor's remarks were clearly intended to prepare the legislature to accept a new legal framework in sharp contrast to established practice. Judge Howell's mining code was of general applicability throughout the territory and its organization; terminology and scope in many significant aspects paralleled the Spanish *Ordenanzas de Mineria* promulgated by Charles III of Spain on May 22, 1783.

This code dictating every step of initiation, perfection, maintenance, and transfer of mining claims upset the virtual complete freedom the miners had enjoyed within their self-declared mining districts where it was clear that they could change the rules with very little effort and disputes were required to be settled by a jury of miners presided over by their own judge. The comprehensive code was also in marked contrast with actions taken in other western territories at the time. Montana, for example, in an act approved December 26, 1864, modeled a twelve-section code after the typical mining district regulations then in force, while the legislatures did not attempt to define rights but under their civil practice act permitted "proof of the customs, usages or regulations established and in force in the mining districts, embracing such claims and such customs, usages or regulations, when not in conflict with the laws of the Territory, shall govern the decision of the action."[7]

The mining code was initially referred to a house committee on October 4, 1864,[8] and not surprisingly, the proposal brought immediate opposition from the miners around the Prescott area. Accordingly, the Journals of the House of Representatives indicate that on October 19, 1864, "Mr. Giles presented remonstrances from miners of Hassayampa, Walker, Turkey Creek, Quartz Mountain and other mining districts, protesting against the passage of a general mining law, which were received, read and referred to the Committee on Mines."[9] Thus, the clear view of many of the individual miners was that no direction at all was needed from government and that the legislature should have simply recognized the right of miners to self-government through the establishment of mining districts that, in turn, would be free to regulate property rights and settle disputes through their own regulations.

The miner's objections fell on deaf ears as the house approved the law as a committee report without objection on October 27, 1864, as substitute House Bill No. 5.[10] During deliberations on the bill, on October 31, 1864, by House Concurrent Resolution No. 6,[11] the eventual mining code was ordered to be translated into Spanish along with the civil and criminal codes, giving a clear indication that the Mexican population of southern Arizona would not be subjected to the same discrimination that had been previously prevalent in the operation of the mines near Prescott.[12] It seems apparent that the legislators were willing to accede to Judge Howell's judgment over the Prescott miners, as on November 3, 1864, the Act for the Regulation of Mines, as amended by the council, passed the council on a six to one vote (Mark Aldrich of Tucson casting the nay vote) and without any opposition in the house. This influence must have been significant considering that thirteen of the twenty-seven members of the legislative assembly were either miners or mining engineers.[13] It is unclear what, if any, lobbying was done for the mining code, but it is possible that the assembly was influenced by Coles Bashford's views and a feeling that the silver mines of southern Arizona were more likely to begin production if the new law more nearly paralleled the district regulations of the Cerro Colorado Mining District and the *Ordenanzas de Mineria* than the regulations of the mining districts around Prescott. Judge Howell also reflected a bias in favor of southern Arizona in his comment that the mines of Pima County were the ones most likely to produce any revenues because of Native American hostilities in the northern part of the territory. He wrote: "Now for minerals: I have traveled five hundred miles in the Territory, and been fifty days on the road. While Indian hostilities continue, the silver mines of the southern portion are much surer and safer than any other, as we have sufficient protection to work them. Every day further work develops their richness and extent."[14]

The session ended seven days later with the comprehensive mining code contained in the Howell Code as chapter 50, containing fifty-two sections and covering fourteen pages of the original text.[15] In spite of the objections to the territorial code, the miners did not have long to make adjustments.

The initial provision of the Howell Code specified that the mining law applied to public lands as well as minerals discovered on the lands of private individuals. This supremacy of the mineral estate is no more obviously stated than in the territorial bill of rights: "The precious metals are the jewels of sovereignty and adheres in the supreme sovereign power. No person can acquire title to any such precious metals without the express consent of such power."[16] This regal ring, a paraphrase of the *Ordenanzas de Mineria*,[17] remained the law of Arizona until its tone was administratively removed in the 1901 Revised Code apparently by authority granted to the Code Commission to "eliminate therefrom all crude, imperfect and contradictory matter." The amended version read: "No person can acquire absolute title to any public domain in which precious metals may be found without the express consent of the United States."

Title I of the territorial mining code contained the general administrative provisions, which initially vested the county probate judges with jurisdiction over mines and set up a record system consisting of three books of mining records that were designated as A, B, and C.[18] Book A was to contain the claims existing on the date of the act, book B was for the initial registration of new claims, and book C contained a second registration of claims required after the completion of the required acts of location set forth in the further provisions of the code. Book C also contained the charters of mining companies and copartnerships. These general provisions showed a desire to keep the records both current and to avoid any question of ownership, which was emphasized by a further requirement that all claims be registered in the real names of the parties, and that any sales had to be recorded within three months to avoid a forfeiture.[19]

Title II of the mining code contained the procedures for initiating, perfecting, and maintaining the mining claims. The law recognized both a mining claim and "auxiliary lands" for surface uses in connection with the mining claim itself. The mining claim itself was called a *pertenencia*,[20] using the Spanish terminology of the 1783 code, and was defined as being two hundred yards square including the vein or mineral deposits and "following the dip of the vein so far as it can or may be worked with all earth and mineral therein."[21] The mining districts were given leeway to prescribe smaller dimensions within the individual districts.[22] "Auxiliary lands," that is, those lands not containing minerals but which were needed for proper mining of the deposit, could be taken up as a part of the location process and could contain up to 160 acres.

The law applied only to lode deposits. Placer mining, although recognized, was given short shrift by the following statement: "The extraction of gold from alluvial and diluvial deposits, generally termed placer mining, shall not be considered mining proper, and shall not entitle persons occupied in it to the provisions of this chapter, nor shall any previous section of this chapter be so construed as to refer to the extraction of gold from the above-mentioned deposits."[23] Thus, placer miners held rights by possession only and presumably by rights defined within the mining district regulations.

The boundaries of mining claims had to be marked with substantial monuments and were required to be witnessed either by the recorder of the mining district (who was the clerk of the probate judge) or any other witness who could prove the proper monumenting to the recorder's satisfaction. The notice of claim had to be posted at the opening of the vein and entered into the recorder's book within three months from the initial act of location.[24] Boundaries could be changed so long as it could be done without prejudice to other parties, but did have to have the sanction of the probate judge, and new monuments had to be fixed at once upon removal of the originals.[25]

Title IV of the mining code provided a means of funding educational institutions in the territory, and by making such a provision Howell showed an awareness of a federal congressional practice of setting aside land for the benefit of common schools at the time territories were created, lands that were granted to the new state upon its admission.[26] Thus, as a part of the process of defining a new claim, the discoverer was required to lay off and define the boundaries of one *pertenencia* (that is, a six-hundred-foot square) adjoining the new claim, which was the property of the Territory of Arizona. The territorial claim had to be presented to the recorder at the time of the initial location, and failure to do so would result in the forfeiture of the claim and any discovery. Once the territorial claim was initiated, the locator had no further obligation to perform work on the territory's claim, and it was not considered to be abandoned so long as it was the property of the territory. The territory could, however, sell the claim, in which case the time within which the purchaser would be required to work the claim commenced upon the date of sale. The territory, likewise, had the right to sell these claims at auction to the highest bidder upon public notice of the same. The amount of the purchase price went into a fund to protect the people of the territory against hostilities, and after "all hostile Indian Tribes in this Territory are liquidated," then all remaining funds would be applied to a sinking fund for school purposes.[27]

A common problem in the early mining districts occurred when two or more persons attempted to claim the same ground. In California, as early as 1860, the courts had recognized a bare right of possession of the first claimant against later competitors.[28] This right of *pedis possessio* has never been clearly defined and has been fertile ground for litigation ever since.[29] The mining code, calling on the wisdom of Spanish legal experience,[30] specified that when two or more parties were exploring the same vein at different places and without the knowledge of the other, the party able to prove the first occupancy had the right of first location. The situs of the first claim was fixed by using the primary excavation as the center of the claim and measuring off the remainder of the claim along the general direction of the vein or deposit. The other parties then could proceed after the first parties had fixed their boundaries. Then, if vacant ground remained between the parties, the claimant making the first discovery had the option of changing the boundaries to take in the vacant ground to its maximum advantage. All latecomers had to locate in the order of the time of their arrival on the vein or mineral deposit.[31] Where one party knew another party was exploring the same claim, and if the first party failed to give notice of the prior claim either verbally or in writing, then the portion of the "mine" situated between the main excavations of the two parties was divided equally between them, without regard to the number of members in each company.[32]

If, in the construction of tunnels or any subterranean works, new veins or deposits were encountered in ground not claimed or owned by the parties, they became the property of the party for whom the tunnel was being constructed and could be claimed in accordance with the normal procedures.[33]

Another historic conflict area has been between the mineral claimant and the occupant of the surface. Because the mining code recognized the right of miners to claim discoveries on private lands, the possible conflict between the surface and mineral estate was foreseen, and the code defined the rights of the respective rights of the miner and the surface owner. The code thus provided that the miner, before sinking any shaft on private land, was required to pay compensation to the surface owner. In the absence of an agreement, the amount of damages was fixed by the probate judge, and in the absence of an agreement and prior to the determination of damages, either party could require the judge to fix the amount of a bond to permit work to proceed. The party responsible for damages arising from subterranean works was the party for whose benefit the tunneling was done.[34]

In addition to the rights of the surface owner, and since tunnels could be made for the purpose of drainage, ventilation, or better handling of ores, it was lawful for any party to construct such tunnel or drift through all private and public property, and the mining code also fixed the rights of owners of other mineral rights. Thus, the code provided, if the tunneling traversed or intersected mineral deposits or ran along lodes claimed and held by other parties, the owners of the other mineral deposit had the option either to pay one-half of the expense of excavation for the distance that the tunnel ran through their mineral deposit and thereby obtain all of the ores extracted from within such length of the tunnel, or to pay nothing and receive half of the ores within the tunnel.[35]

Mining claims were perfected by requiring the claimant during the first year after location to sink at least one shaft to a depth of thirty feet or to run a tunnel of fifty feet in length into the main body of the vein or the adjoining rock. Upon completion of this work, the recorder had to be notified, and in turn examined the work, made a record and certificate of the result of the examination, and obtained three specimens taken from different parts of the work. The specimens were kept in the office of the probate court to be preserved for the use "of the mineralogical professorship of the University of Arizona."[36]

Upon completion of the certification, the code permitted the claimant, upon petition to the probate judge, to request a confirmation title. Upon such request, the judge was required to issue a summons requiring all interested persons to appear in court within sixty days from the notice to show cause why the title of the claimant should not be confirmed. This notice was published twice in the territorial newspaper and was posted in the probate clerk's office. Thereafter, upon a finding that claimant had complied with the provisions of the law, the probate judge could enter a decree granting a "perfect title" to the claim until the first day of January 1868 (recognizing a "holiday" from the performance of work commitments during this period), "and forever after unless abandoned by them." If a dispute over title arose in this process, an answer could be filed by the return date of the summons, in which case the question of possessory rights was settled pursuant to the normal rules of civil procedure.[37]

A two-year time frame was allowed after the conferring of the initial title as a grace period to allow the claimant to develop the mine and to procure machinery to work the claim. Claimants were, however, required to file annual affidavits with the probate clerk prior to June 1 of each year stating that they had not abandoned the claims but intended to work them in good faith. After the expiration of the two-year term, it was obligatory upon the claimants to "hold actual possession of them, and work the vein," which was considered as being thirty days' work per year.[38] In the case of auxiliary lands or millsites, the requirement was made that the claimant had three years to spend $100 in the construction of a mill and thereafter had to work ore for thirty days per year.[39]

Where one person located a number of claims on the same vein or a number of separate individuals having claims on the same vein agreed to work together, they could concentrate labor, capital, and energy at a single point that was deemed to be "best suited to ascertain the best advantage of the general character, quality, and capacity of that particular vein or mineral deposit," and then proceed under the similar provisions of the code.[40] This provision may have been an attempt to encourage joint development so that outside financing would not be needed. Judge Howell's writings make it clear that he felt that existing financial practices lent themselves to abuse. He at one time observed that the solution to the problem of lack of mineral production was "to get a local resident to locate the mine and begin work," and above all else "keep out of Wall Street, as the fountain head can be reached much cheaper through the right channel."[41]

Both the requirement for the performance of the initial location work and the performance of the annual work could be suspended if the claimants were prevented from working the vein because of the hostility of Indians or other good cause rendering the work difficult or dangerous. This authorization for suspension of annual work had to be through the probate judge who would not grant more than one year at a time during the continuance of the cause.[42]

The abandonment and relocation of mines and mining rights was covered in title III of the mining code and provided another direct link to the Spanish laws through the use of the term "denouncement" for the process of relocation of abandoned claims. These provisions initially specified that any claimant who did not comply with the provisions of the mining code would forfeit all right to such recorded or unrecorded claims and auxiliary tracts and could not refile on such claims within a period of three years after the forfeiture.[43] Thus, the practice of making repeated filings to avoid the obligation to dig a shaft or run a tunnel was prevented.

The denouncement provisions applied to all veins and mineral deposits situated on public lands that had not been worked and occupied from the time of acquisition of the territory by the United States. Thus, any claims or other rights claiming the authority of Mexican law were considered to be abandoned, and such claims would be subject to relocation as provided for new claims. The act of relocation or denouncement required the relocator,[44] after the initial registration of the claim, to give the former owners notice of the relocation and permit such owners three months within which to remove anything of value from the claim. This notice was required to be published in the nearest newspaper and by posting "at three of the most conspicuous places in the county where the mine is situated." Three months after expiration of the notice any "buildings, furnaces, arrastras, metals, and every other species of property" that remained on the grounds of the mine became the undisputed property of the new claimant without compensation of any kind to the former owner or any other party. Presumably, such an action would, in addition to cutting off the rights of the former owner, likewise sever any rights of creditors. If the person taking possession entered upon a mining claim before it was abandoned, however, the purported relocator could be ousted "in a summary manner" by order of the probate judge and required to pay damages and the cost of such proceedings.

The existing self-government within the mining district and their future operation was formalized and organized in three sections of the mining code. Initially, the validity of all laws and proceedings of mining districts prior to the date of the code was recognized within certain limitations.[45] First, no perpetual right was recognized, and permanent rights could be established only in accordance with procedures set forth in the mining code. All claimants under prior law had six months within which to file their claim with the probate clerk. Further, the recording officer of each mining district was required to deposit the district's records with the probate clerk within three months after the effective date of the Howell Code, and such records could thereafter be received as evidence in any courts of the territory regardless of any defects in form so long as their contents could be understood.[46] The same three-month limitation was also required for the recordation of all conveyances of mining claims, all records of which were to be maintained by the probate clerk.[47]

The records held by the district probate clerk were required to be open to public inspection during normal office hours, and the clerk was required to make certified copies that were admissible in the courts of justice. The failure of the recorder, registrar, or other recording officer of each of the mining districts to comply with the provisions for filing was punishable as an act of contempt by the probate judge and by a fine not exceeding $5,000 and imprisonment of not more than a year, and, further, such official would be incapable of holding such office and any mining claim. This was, undoubtedly, a very severe penalty in those days.

The authority for the continuing viability of mining districts first required the filing of the records of the district with the clerk of the probate court,[48] at which time the mining district recorder was required to take an oath before the probate judge that he would faithfully perform his duties until his successor was elected and qualified. These duties were defined as being available upon the request of a claimant to go upon any mining claims to measure the boundary markings and description of the location notice and to certify that the procedures for making the location had been properly taken. This certification could be made either through personal observations or by evidence of a credible witness who was present when the same was done. If a witness was used, the witness's name was required to be listed on the record. A mining district recorder was also required, upon the request of a claimant, to examine the shaft or tunnel and to make the appropriate measurements and issue a certification of completion. The recorder was required to make quarterly reports to the probate court to which was attached a certified copy of all records made by him during the preceding quarter. The district recorder was also authorized to accept the appropriate fees and to turn these fees over to the clerk of the probate court apparently at the time of the quarterly report. These fees were fixed in section 9 of the mining code, which required a $1 fee for registration of the claims, most notices, and all conveyances. A $2 fee was required for the final registration of the claim in book C, and the registration of a mining partnership required a fee of $2.50. The fee paid to the recorder was apparently not taken out of these recording fees, but instead the districts would fix the compensation for the recorders.

New mining districts could be formed by twelve or more people owning mining claims within any mining district, a contiguous district, or in any area not included within an established district. The initial act of creating a new district was through the publication of a notice specifying the limits of the contemplated district and signed by the petitioners for the new district. This notice was posted in three conspicuous places within the district, and if a part of an established district was included, a copy of the notice was required to be left with the recorder of the mining district at least ten days before the date of the meeting. At the meeting, all claimants within the proposed boundaries could vote on the establishment of the new district and its limits, so long as they were within the boundaries outlined in the notice. If the new district were approved, the meeting would proceed to select a name, establish laws, and elect a recorder who would thereafter perform the duties in accordance with the mining code and file the proceedings of the meeting with the clerk of the probate court.

Title IV of the mining code provided eight sections containing the procedure for litigation of mining cases. These proceedings were governed according to the rules of equity based on the merits of each party's position and apparently recognizing the fact that legal title was still open to question. The probate court was the trial court, and all appeals were to be taken to the district court within ten days after judgment, provided that the appellant first posted a bond sufficient to cover all costs and damages that might be suffered by the adverse party. Any appeal of the trial court proceedings was taken to the district court and was considered solely on the record and testimony from the probate court, provided, however, that new testimony could be introduced if it could be shown that "new, important, or material testimony has been discovered since the trial before the probate court, or could not be procured in season to be used at said trial, and which may change the judgment in the cause." In such a case the new testimony was to be taken by the judge, clerk, or commissioner, and became part of the record.[49] The district courts also exercised supervisory control and the review of the actions of the probate court through mandamus, prohibition, and injunction. Any questions of errors of law could also be considered by the district court and further appeal allowed to the Territorial Supreme Court.

Because of the miners' long-standing distrust of the civilian judicial system, it is interesting to note that a statutory provision was made recognizing the need for expertise and arbitration of mining disputes. The probate judge was authorized, "when local knowledge is necessary" either at his own option or at the request of either party, to appoint a commission of three persons "skilled in mining" to examine the question at issue and make a report in writing and under oath of all the facts and circumstances bearing upon the litigation.[50] This report could either be rejected or approved in whole or in part by the probate judge and would become part of the record. These commissioners received five dollars per day for their services; these costs were part of the cost of litigation. Boundary disputes (i.e., lines, divisions and demarcations) between parties claiming mineral veins, deposits, or auxiliary lands were determined by the probate court in a summary manner, or, upon the request of either party, by an appointed commission of miners.

Disputes between owners of undivided interests in mineral deposits have historically been a problem, and the common law has traditionally provided scant authority to resolve questions of partition of mineral estates and rights between uncooperative co-owners seeking to develop the same land.[51] The mining code made detailed provision to resolve both of these problems.[52] The code thus provided that whenever co-owners of mineral lands could not agree as to the working or management, one or more of them could file a complaint with the judge of the probate court setting forth the facts, and the judge would thereupon issue a summons to the other members or co-owners requiring them to answer the complaint. If, on hearing the complaint, the judge was of the opinion that the claims were susceptible of partition and could be equitably made, and if the parties could not agree among themselves on making the partition, the judge was required to appoint the mining commission to go upon the claims and segregate the claims "having in view the equities of the case and the customs and regulations of the mines." Upon giving the report to the judge, the judge would enter a decree in accordance with the report, which, unless appealed, was conclusive between the parties. Where one or more partners owned not less than one-third of the full interest of the mine or the partnership and desired to work the mine and the other parties refused to join in the effort, the party desiring to initiate work could also file a complaint with the probate judge. If the judge, on hearing the complaint, believed that the complainants were able to work the mine and that there was no good reason why the mine should not be worked, the judge was required to enter a decree to that effect. If a co-owner did not appear at the proceedings, the judge could appoint an agent or receiver to protect the rights of the absent owners to see that the mine was properly worked and to receive their proportion of the profits. However, the owners working the mine by any such authorization were required to conduct all the work at their own expense and were not authorized to bind the absent owners to payment of any indebtedness that they might incur on their behalf.

The United States Congress Finally Acts

On July 26, 1866, a bare eighteen months after the effective date of the territorial mining code, under the rather deceiving title of "An Act Granting the Right-of-Way to Ditch and Canal Owners over the Public Lands, and for Other Purposes,"[53] the first general law governing the disposition of mineral rights in the public lands of the United States was passed by the United States Congress.

Surprisingly, Congress had been extremely resistant to take any action and attempt to control the western mineral resources immediately after the 1848 discovery in California. Perhaps the self-governing structure of the mining camps appealed to the political philosophy of the day, or the nation was simply too preoccupied with the slavery issue and the threat of secession by the Southern states. In any case, the reluctance was real and showed in remarks made by President Fillmore on December 2, 1851: "I am inclined . . . to advise that [the mines] be permitted to remain as at present, a common field, open to the enterprise and industry of all our citizens, until further experience shall have developed the best policy to be ultimately adopted."[54] This lack of a congressional mandate clarifying the rights of individuals to mineral lands of the United States only served to add credibility, if not official sanction, to lawmaking efforts of the miners in the mining districts in the California goldfields.

As the 1860s approached, the inevitable conflict between the Northern and Southern factions of the United States, the obvious requirement for funds, and foreign pressure for land tenure to justify investment in the California goldfields again stirred the United States Congress to attempt to do something about the mines. Racial issues continued to be the stumbling block, and when, on April 11, 1860, Senator Gwin of California introduced a bill as an amendment to the Homestead Act to retroactively legalize mining in California and Oregon by anyone who had declared their intention to become a citizen,[55] the motivation for the bill, as reflected in the debates, was to prevent mining by the fifty thousand to sixty thousand Chinese "locusts and grasshoppers" in California.

This bill also passed the Senate and again died in the House. During the debates, the principal dissenting voice was that of Thomas Ewing, a former secretary of the interior and then a senator from Ohio, who continued to push proposals he had originally made in 1849. Senator Ewing was not the only voice suggesting that the Congress should look to several hundred years of practice of the Spanish in Mexico. Herman Ehrenberg, the president of the Sonora Exploring & Mining Company in Arizona Territory, in a letter to William P. Blake, the editor of *The Mining Magazine and*

Journal of Geology, published in December 1859, suggested that the Congress enact a system based on the 1783 Spanish mining ordinances.

Though a similar format became the Arizona territorial mining law in 1864, Ehrenberg's proposals were intended as a federal mining law with a clear intent to permit the development of the mines and defining the miners, rights, privileges, and duties, thereby preserving title to any mineral deposit "discovered by enterprising explorers."[56] Professor Blake, in adding an editor's comment at the end of Ehrenberg's letter, expressed his view that the mining law was a subject of national importance and "the foundation of the prosperity of many enterprises in the Territories." He went on to state that "capitalists [because of the lack of law] were not willing to expend their money in sinking costly shafts, and erecting machinery and buildings, to prove a vein which by miner's law, might be indefinitely subdivided according to its richness."[57]

Fig. 6. Sylvester Mowry (1833–71). The owner of the Mowry Silver Mine and the target of the Union column, Mowry believed that he would provide silver and lead to Confederate forces. He is best remembered as an early advocate for establishing the Arizona Territory, having published materials in support of this effort. He was arrested as a traitor by Union forces and eventually died in England. Courtesy of the Arizona Historical Society, #18136.

The debate in the United States Congress over the mining laws was lost in the preoccupation with the early years of the Civil War but began in earnest in 1864. One of the first suggestions, introduced in the form of a resolution in the House of Representatives, would have authorized the seizure of the mines of Colorado and Arizona, which would thereafter be sold to pay the debts incurred by the Civil War. The miners were, predictably, furious. Sylvester Mowry, who in 1856 had purchased a silver mine in the Patagonia Mountains of New Mexico Territory (now Arizona), and who was, incidentally, an outspoken Confederate sympathizer,[58] wrote an angry letter published in the *New York World* of April 25, 1864. He pointed out that a mine was not a piece of mineralized ground but only "had value according to the amount of labor and capital employed in demonstrating its extent and capacity of production,"[59] and suggested that a taking, "if it could be enforced, would at once put an end to our existing civilization, and the great American desert, which has been made, in spite of governmental negligence and worse, to 'blossom as the rose,' would return to its pristine worthlessness."[60]

He argued that the economy of the California goldfields, the silver mines of the Comstock, and the mines of Colorado had flourished in spite of the government and to now assert that "you are trespassers on the public domain" would have to be enforced by force.[61]

Mowry's argument was simple: "In the infancy (magnificent thought [*sic*] it is) of our mineral development no better precedent can be followed than the wisdom of the Spanish law. Give the miners title to their mines, and impose a fair tax. It will be paid readily and honestly. If it is made onerous it will impede the opening of new mines, and thus 'kill the goose with the golden egg.'"[62] He finally concluded that any seizure of the mines from their "actual *bona fide* possessors" would result in either a forcible stoppage of all production or a revolt by "the people of the frontier that no army can put down."[63]

Another debate of the 37th Congress concerned the potential revenues from taxation of the mines. Mowry also had opinions on this subject and echoed the prevailing view of the miners in a letter to the *New York Herald* on May 2, 1864. He initially pointed out that "if a tax is laid on the gross proceeds of the mines, it is not at once apparent that you impede, if not absolutely stop, the further development of non-paying mines."[64]

In view of the immense risk of financial failure (and physical danger), Mowry argued, such a tax "amounts simply to taxing a man for what he has not got; and, worse than that, to put to death all 'prospecting,' and to stop at once every mine that did not yield an enormous profit."[65] The suggestion that the mines should pay a special tax because they were on public lands was summarily dismissed by Mowry as the same as an equal tax on grains and livestock of the settler on the public lands. He suggested that if there were to be such discrimination it "should be against the farmer and grazier, as his work is light, and his capital nothing compared with that of the miner, while his profits are more certain."[66]

Mowry had especially strong words for a senator from Michigan who proposed a 5 percent tax on the gross yield of the mines, and after pointing to tariffs protecting cotton and iron, and bounties for the cod fisheries: "It is now proposed, in the very death-agony of the republic, to strike a fatal blow, by unjust and suicidal taxation, at the very greatest hope of the country, viz, the protection of the precious metals? The wisest thing that the present Congress can do—the only thing it can do of service to the government and justice to the miner—is to impose a fair tax on the net proceeds of the mines."[67] In the 39th Congress of the United States, however, the sentiment continued to be strong to sell all the mines, including the mines of California, to obtain revenue. The first bill was introduced in both houses by Sen. John Sherman of Ohio and Rep. George W. Julian of Indiana. Senator Sherman (who would later become secretary of the treasury during the Rutherford Hayes administration, 1877–81) may have been influenced in his efforts by the opinions of former interior secretary Thomas Ewing, who had been a longtime close family friend and who had adopted his brother, Gen. William Tecumseh Sherman, after the death of his father, and General Sherman himself, who was very familiar with the California goldfields as a result of his assignment to Colonel Mason's staff and who had thereafter set himself up as a banker in California during the 1850s. This initial proposal was strongly supported by the secretary of the treasury and drew the battle lines for the passage of a mining law.

The principal antagonists in this legislative battle turned out to be Representative Julian, the chairman of the House Committee on Public Lands, and Sen. William M. Stewart of Nevada, a member of the Senate Committee on Mines and Minerals. Representative Julian was the champion of efforts to sell the mineral lands, and in both 1865 and 1866 introduced legislation to accomplish this end. The sales legislation continued to be opposed by the western delegation led by Sen. John Conness of California and Senator Stewart. Senator Stewart's plan was to obtain formal recognition of the past practices of the miners and proved to be the most enthusiastic and persuasive legislator in his efforts. These efforts began in 1865 when Senator Stewart, in an attempt to overturn a Supreme Court decision holding that the federal courts had no appellate jurisdiction to review state court decisions involving contests between private claimants on land owned by the United States,[68] introduced and was successful in having passed legislation wherein actions for recovery of mining titles would not be affected by the fact that the land was owned by the United States but that "each case shall be adjudged upon the law of Possession."[69] Thus, in 1865, the federal law recognized by implication that there was indeed *a law of possession* that at this time remained uncodified except by the mining district regulations.

The events in Washington were closely monitored by the western mining communities, and the newspapers in Nevada and California were particularly vocal in imploring the Congress not to destroy the western mining economy that they believed to be premised on the efforts of the individual miner under the flexible practices of the mining districts and to otherwise make sure that as much land as possible was available for mining purposes.[70] The attitude of the miners in southern Arizona was somewhat different, however, and Ehrenberg's proposals made six years earlier to Professor Blake providing for some administrative oversight of the claim-staking process, mining claims of uniform size, and vertical boundaries was under consideration in 1865 by the House Committee on Public Lands. This consideration was brief, however, and the only impression left was a single resolution: "Resolved, that the Committee on Public Lands be, and they are hereby, instructed to inquire into the expediency of adopting the code of mining laws passed by the legislative assembly of the Territory of Arizona, hereto appended."[71]

Richard McCormick, having now been elevated from territorial secretary to governor, also reflected the fear, as late as a month before the law was enacted, that mineral lands would be sold, and weighed in with a letter published in the *Evening Post* in New York:

> The action of Congress in reference to mineral lands is keenly watched by our people without exception they are opposed to a sale of the lands, believing that it will lead to a monopoly and speculation. It will require all the wisdom of our national legislators, if the present laws are changed, to devise a plan whereby the people will be satisfied and the best interest of the country consulted that the nation has

> a right to look to its mineral regions for a larger return. It is not questioned, but that such return is to be had by a sale of the lands is not yet apparent. Whatever legislation is at, it fit the generous in its dealing with the hearty and adventuresome men who have risked their lives and encountered untold privations in opening and occupying these remote wilds.[72]

As it turned out, the most significant event of the prior year was the creation of committees on mines and mining in both the House and Senate, committees that figured prominently in the enactment of a mining law during 1866. The year of 1865 began with a good omen for the miners as Senator Stewart was able to shepherd through the Congress legislation to confirm possessory rights to mining claims acquired in compliance with rules and regulations in Nevada and to grant Adolph Sutro a fee title to certain mineral lands to allow him to construct the famous Sutro Tunnel that was to drain the lower workings of the Comstock Lode.[73]

When it came to the mining law itself, the echo of the position taken by Senator Benton in 1849 continued to be strong, and undoubtedly many remembered his admonishment that although the California Gold Rush required no fee simple title because of the nature of the temporary values to be recovered by washings, rights required by industry were different. Indeed: "Mining requires great capital—a fee simple estate—and a large tract of land—and is worked for an age or centuries. After these washings are exhausted, mining may follow; and that is the time to sell the fee simple, but not in patches of 2 acres. To the washer 2 acres is a 100 times too much; to the miner it is a thousand times too little."

In April 1866 a bill was introduced by Senator Sherman, having apparently now changed his mind from his earlier insistence on sales legislation, to regulate the occupation of and extend preemption rights to mineral land. The Committee on Mines and Minerals amended the bill, chiefly under the authorship of Senator Stewart, opening mineral lands to free exploitation and permitting acquisition of fee simple title for a nominal amount for lode mining claims. Senator Stewart's argument in favor of the bill was that by providing secure mining titles, investments would be encouraged, resulting in an increased production of gold. Since the principal asset of the United States backing up his debts was land, the increased production of gold would proportionately raise the value of land and thus the debt would be proportionately reduced. Commentators have quite correctly noted that

Fig. 7. William M. Stewart (1827–1909). Stewart served as U.S. senator from California and was principally responsible for the passage of the first federal mining law in 1866. He was also reputed to have been the author of several of the lode mining district regulations in California. Courtesy of the Nevada Historical Society.

this logic does not seem altogether clear in the modern context, although inflation from gold production may have reduced the real cost of the debt, it is equally apparent that the main concern of the western representatives and their rather loud constituencies was a clamor for security of title and the ability to purchase mineral lands at a reasonable price. Whatever argument was used in the halls of Congress therefore seems beside the point. The bill passed the Senate, and upon its arrival in the House, Representative Julian had the bill referred to his committee on public lands—where he obviously intended it would die.

The mining delegation, in mounting a new offensive of legislative manipulation, discovered that a bill authorizing rights-of-way over public lands in California had been submitted to Representative Julian's committee by mistake. The mining delegation had this mistake corrected and transferred the bill to the Committee on Mines and Mining. After the bill was reported out of the committee and passed by the House, the Senate committee, by amendment, substituted its entire substance with the original provisions of Senator Stewart's bill that had been passed three weeks earlier. When the legislation came back to the House, it was referred to the committee from which it came, that is, the House Committee on Mines and Mining, and thereby avoided Representative Julian's committee. The bill sailed through the House with very little opposition, although Representative Julian loudly complained of this "trickery." Thus, the new lode mining law became law on July 26, 1866, under the rather deceiving title of "An Act Granting the Right-of-Way to Ditch and Canal Owners over the Public Lands, and for Other Purposes." This is not to suggest that the legislative maneuvering resulted in the passage of a bill that was not the will of Congress—the bill was fully debated, and the 1866 act was clearly the will of Congress.

The Federal Mining Laws, 1866–72

The new federal law was short, containing eleven sections, only four of which directly impacted the procedures for acquiring mineral title. In fact, the substance of the bill came directly from the lode mining regulations of the Grass Valley Mining District of Nevada County, California, adopted on December 20, 1852, and the Gold Mountain Mining District of Storey County, Nevada, adopted on March 4, 1860, both of which were shaped in some degree by Senator Stewart.[74] By the act, the mineral lands were declared to be open to exploration by all citizens of the United States and those who had declared their intentions to become citizens subject both to regulations as may be prescribed by law and "local customs or rules of miners in the several mining districts" insofar as they were not in conflict with the laws of the United States. After any person or association of persons had expended $1,000 in actual labor and improvements on a claim, the claimant could obtain a patent (fee title) for the vein contained within the claim together with the right "to follow such vein or lode, with its dips, angles and variations to any depth" even though it entered adjoining land. The patent was limited to the single vein for which it was granted, and each locator was limited to 200 feet in length along the vein—with the exception of the original discoverer, who was entitled to an additional claim in recognition of the discovery. Individuals were prohibited from having more than one location on the same lode, and not more than 3,000 feet could be taken in any one claim by any association of persons. The individual states or territories were given authority to provide rules for the working of mines within their borders, including easement and drainage, and local courts were given authority to determine contests of title prior to the issuance of patent.

This law was recognized by everyone, including the miners, as being far from perfect, with sentiments running all the way from those who still felt that the sale of the mines was the appropriate method of disposal to others who felt that the mining law itself was an infringement upon established rights.[75] Gregory Yale put it this way:

> As the initial act to the legislation which must necessarily follow, it is more commendable as an acknowledgment of the justice and necessity which dictated it, and for its expediency as a means to the advancement of the State and nation than for the perfection of its provisions, or their exact adaption to the accomplishment of the object intended. We must not, however, find fault with the law on account of its imperfections, or the introduction of objectionable feature in the mode to be followed in acquiring a title under it. These imperfections can be remedied, the rights of the parties amplified in many particulars, and the system so changed as to work with more facility than now anticipated.[76]

In answering those elements of the mining community critical of the law, Yale concluded:

> The result of the whole fight is the grant of all mines to the miners, with some wholesome regulations as to the manner of holding and working them, which are not in conflict with existing mining laws, but simply give uniformity and consistency to the whole system. The escape from entire confiscation was much more narrow than the good people of California ever supposed. If either of the bills originally introduced had been passed, the Pacific states and Territories would have received a blow from which they would never have recovered. The Government could only have receded after the most irreparable and widespread damage had been done.[77]

The 1866 law, however, contained significant flaws. One of the most difficult aspects of the law was that it applied only to vein-type deposits, the patent issued by the government was only for the single vein within the claim, and the extent of surface ground fixed by the patent was undetermined. This resulted in "piratical" locations made within the immediate vicinity of bona fide claims, which had the effect of preventing the complete development of the major vein. The ultimate effect of this was seen as potential issuance of patents that would cut up valuable property into worthless fragments.[78] Another deficiency of the 1866 law was that no provisions were made to provide for uniform assessment or maintenance work on the claims (other than the $1,000 required for patent). The apparent concept of the framers was that assessment work requirement would be mandated by district regulations. However, commentators pointed out that district regulations were volatile, and a loss of title could result where rules were changed during an individual miner's absence.[79]

The miners, however, recognizing that the new law was the product of their own rules, looked upon all attempts to enact a more complete code with distrust as they feared "that to open the subject again might imperil advantages already gained . . ."[80] The core of the principles that the miners wished to retain were the "recognition of their mining laws, and the right of the discoverer of a mine to purchase the title from the government at a reasonable price."[81]

The miners had little reason to fear. The United States was then emerging from the Civil War, relations with Great Britain were perilous at best, and fears were being expressed in the Congress as to whether the nation could retain the Pacific territories acquired in the war with Mexico and the earlier treaty with Great Britain.[82] In fact, the rapid development of the mines fueled an enormous growth in the western United States, thereby confirming the philosophy of the proponents of the political theories of manifest destiny; the mining lobby thus had little trouble in accomplishing some major modifications of the law without disturbing the basic principles.

The mining law was then a major subject of debate over the ensuing six years. In 1870, the law was extended to allow the purchase of placer ground and then combined with the earlier 1866 law to create the General Mining Law passed by the Congress on May 10, 1872. In doing so, Congress did not change the basic principles of the law, but it was much more than a simple reenactment and combination of the Lode and Placer Acts and did significantly more than "oiling the machinery a little"—as it was characterized by its principal author, Rep. A. A. Sargent of California.[83]

The core of the miners' concerns was preserved by the provisions permitting the issuance of patents for mining claims at $2.50 per acre for placer ground and $5 an acre for lodes. This was considered a reasonable price by the miners in 1872, although when viewed in context of the standard price for agricultural lands at $1.25 an acre it could not have been considered an insignificant amount.

The law fixed the size of lode claims at 1,500 feet along the apex of the lode at the surface and 300 feet on either side of the centerline and established the size of placer claims to 20 acres while permitting a group of not more than eight persons to associate to aggregate a placer claims of 160 acres and while also permitting a claim for either associated or independent millsites of 5 acres and an "exploration" tunnel site that would allow lode claims to be staked based on a discovery of valuable mineral within a tunnel driven for that purpose.[84]

A major concession of the 1872 law, however, most awaited by the miners read:

> The miners of each mining district may make regulations not in conflict with the laws of the United States, or with the laws of the State or Territory in which the district is situated, governing the location, manner of recording, amount of work necessary to hold possession of a mining claim, subject to the following requirements: The location must be distinctly marked on the ground so that its boundaries can be readily traced. All records of mining claims made after May 10, 1872, shall contain the name or names of the locators, the date of the location, and such a description of the claim or claims located by reference to some natural object or permanent monument as will identify the claim.[85]

The Arizona miners were thus given authorization for the continuation of their mining districts but with a more limited scope of authority for what could be included within the regulations.

CHAPTER 4

Revival and Operation of the Arizona Mining District Laws

The Repeal of the Territorial Mining Code

The Arizona Territorial Mining Code, despite its innovations and acclaim (at least by some), never really had the opportunity of being perfected through use and experience because of a basic legal flaw. This flaw was that neither the Territory of Arizona, nor the miners within the several districts, by their regulations, could determine the ultimate legal right of ownership of minerals in the public lands of the United States. This was within the sole purview of the United States Congress. The miners and the various territorial and state legislatures had not ignored this fact but had acted out of the need to fill a vacuum that they always recognized existed. On July 26, 1866, with the enactment of the first federal mining law,[1] the government inaction that provided a legal justification (or at the very least an excuse) for Arizona's territorial mining code was gone.

Word of the new federal mining law reached the people of Arizona through the September 12, 1866, edition of the *Arizona Miner*, where the law was printed in its entirety.[2] It was apparent that the territorial mining code and the federal enactment were not reconcilable, and a call immediately went out in the editorials of the *Miner* for the repeal of Arizona's pioneer effort.[3]

Richard McCormick, having been elevated from territorial secretary to governor, was quick to give his own views to the legislators, and in the Governor's Message to the Third Legislative Assembly printed in the October 13, 1866, edition of the *Arizona Miner*, he praised the territorial mining law as preserving "all that is best in the system created by the miners themselves," with the additional opportunity to obtain permanent title. He elaborated: "It is a more equitable and practicable measure than the people of the mineral districts had supposed Congress would adopt; and credit for its liberal and acceptable provisions is largely due to the influence of the representatives from the Pacific coast, including our own intelligent delegate. While it is not without defects, as a basis of legislation it is highly promising and must lead to stability and method, and so inspire increased confidence and zeal in quartz mining."[4]

The governor then went on to note that the federal law gave authority to the legislatures of any state or territory to provide rules and suggested to the assembly that "it will be your duty to prepare such rules, either by amending the present mining law of the Territory so as to conform to the law of Congress, or by its repeal, and the substitution of an entirely new statute. Whatever your preference in this particular, I would suggest that care be taken to make the required rules as intelligible and comprehensive as possible, and that the recording and preservation of titles, both for the security of the miner and the capitalist and to obviate future litigation, be entrusted only to the most responsible officers."[5]

The issue of work requirements continued to loom and, undoubtedly in a continuing echo of Herman Ehrenberg's criticism of the territorial mining code as enacted, Governor McCormick said:

> It is also important that, excepting in districts where active hostility on the part of the Indian absolutely prevents, the actual occupation and improvement of claims be made a requisite to their possession, unless pre-empted under the Congressional law. The lack of such a requirement hitherto has seriously retarded the development of our mineral resources and the general prosperity of the Territory, and proven discouraging to newcomers, especially in the counties on the Colorado River, where hundreds of lodes, taken up in years past by parties now absent from the Territory, are unworked, and yet, under existing law, no one has a right to lay claim to them, be he ever so able or anxious to open them.[6]

There were some specific proposals for replacement of the territorial code, and the *Daily Arizona Miner* (a daily when the legislature was in session) in its October 26, 1866, edition noted that the congressional law must form the basis of any new territorial law. It was the judgment of W. H. Hardy, a member of the council from Mohave County, among others, that under the federal act, actual possession and improvement of the claim was necessary to hold it, and that every claim must hereafter be held with the view of securing a patent. The new federal law was compared with the preemption laws for patenting private lands, and the miner would be required to initially perform some work to prove possession and then to show that the property was worth developing as a mine.[7]

The calls for change also provide a clue to the frustrations of the miners and some of the problems experienced under the Howell Code. An article in the *Daily Arizona Miner* of October 11, 1866, provides this insight:

> Mr. Hardy has introduced a bill in the Council to repeal Chapter 50 of the Howell Code, which is the mining law of the territory. This is well enough, for without taking into view other reasons, a congressional act makes this long and intricate law unnecessary. But what the people are chiefly interested in is to know the character of the new law. Our legislators propose to substitute. We learn that Dr. Davis of Mohave County has one in hand, and we make no doubt that he will be careful to see that it is short, pointed, and so clear that no man can misunderstand it. A law that is open to a dozen different interpretations is not fitted to this new country, or indeed to any country. Simplicity is a first necessity, and economy is another. Our miners should be able to perfect their titles at the most reasonable costs. Manner of recording is of great importance, as is that of defining the claims, and working them. While the new law is liberal and just, we trust it will be cautious and complete, so that its practical result may be the speedy development of our great mining interest, and it may prove an invitation rather than a source of distrust to the capitalist.[8]

The federal mining law enacted in 1866 had given the miners significant authority to enact their own laws. This grant of legislative authority to an unelected, nongovernmental body was unprecedented, and by section 1 of the law declared that "mineral lands of the public domain" were free and open to exploration and occupation by all citizens of the United States, "subject to such regulations as may be prescribed by law, and subject also to the local customs or rules of miners in the several mining districts, so far as the same may not be in conflict with the laws of the United States."[9]

The Arizona legislature passed on this grant of authority to a new law approved on November 5, 1866, composed of only seven sections, the operative provision of which stated: "The mining districts heretofore created in the several counties of this Territory, are hereby authorized and empowered to make all necessary rules and regulations for the location, registry and working of mines therein; Provided, that all locations and registrations of mines and mineral deposits hereafter made in any of the said districts shall be transmitted to the County Recorder for record within sixty days after the same shall have been located."[10]

The county recorders were authorized to procure suitable books for the records of mines and to receive fees of one dollar for recording each claim so long as it did not exceed one page, in which case each additional page required twenty cents.[11] The claims that had been located in the name of Arizona Territory under the territorial code were declared to be abandoned and open to relocation except those that had been theretofore sold.[12] The law made it clear, however, that placer mining was still not recognized as being under any permanent right and that the act applied only to mineral deposits commonly called veins or lode mines.[13]

Transitional Regulations, 1866–72

With the passage of the first federal mining law in 1866 and the repeal of the Arizona Territorial Mining Code, legislative authority was returned to the mining districts, who in turn resumed many of the same practices that had been in force before the enactment of the territorial code. This was apparently precisely what Congress had envisioned in the 1866 federal act, while giving the miners the right to purchase what was essentially a single mineral vein and limiting the size of lode claims to not more than 200 feet per individual along the vein (with an additional 200 feet for the discoverer) and further limited to a total of 3,000 feet per single claim in the case of an association. The law also granted miners their coveted right "to follow such vein or lode with its dips, angles, and variations to any depth, although it may enter the land adjoining, which land adjoining shall be sold subject to this condition." As for any other requirements, section 1 of the 1866 act specified that the location would be "subject to such regulations as may be prescribed by law, and subject also to the local customs or rules of miners in the several mining districts, so far as the same may not be in conflict with the laws of the United States," and section 2 required that the claim "conform to the local laws, customs and rules of miners."

The regulations thereafter enacted in the various mining districts of the Arizona Territory typically made provisions for only lode claims (even though provisions were added for placer claims under federal law in 1870) and specified the size of the claims to be not more than 200 feet along the course of the vein at the surface and allowed widths of between 50 and 150 feet on either side of the vein.[14] In accordance with the federal law, it was normally also specified that the discoverer was allowed an additional 200 feet along the vein in recognition of the discovery. Of all the regulations enacted during the 1866–72 period, only the Pine Grove Mining District enacted any new provisions for placer claims, which were to be not more than 300 feet long and 100 feet wide.[15]

Perhaps the most comprehensive new regulations enacted during this transitional period were those of the Bradshaw Mountain Mining District near Prescott, created on October 24, 1870. These regulations, prepared by Mack Morris (or McMorris), George Monroe, and Samuel Ball, beginning with article 2, provided:

> Art 2—That this Dist elect a Recorder whose term of office shall be one year from date of election & whose duties shall be to make true & correct entries of notices of mining & all other claims left with him for record in good & suitable books expressly for that purpose, said books of record to always be kept in the possession of the Recorder & subject to inspection to all persons when calling at proper office hours.
>
> Art 3—The Recorder shall receive as compensation for his services the following fees viz—Recording mining & all other claims one dollar for each name & will be required to give a certificate of record to all persons requiring the same, by payment for each certificate one dollar—
>
> Art 4—Any six miners by giving three days notice regularly posed at conspicuous places at different mining camps in the Dis't, can call a regular miners meeting & it will be the duty of the Recorder to act as Secretary of all regularly called miners meetings—
>
> Art 5—No person will be allowed to vote at miners meeting except those owning interest in the Dis't—
>
> Art 6—A claim of any Gold, Silver, Copper, Lead or other metallic quartz or rock on any leade, lode, or ledge, or deposit shall be two hundred feet by pre emption & two hundred feet additional for the right of discovery together with all dips, spurs, angles & variations, and seventy-five feet on each side of the Leade, lode or ledge so claimed, for working & mining purposes—
>
> Art 7—Any person or persons discovering any Leade, lode, ledge or deposit of Gold, Silver, Copper, Lead or Clay[16] & claiming the same

will be required to erect a monument on the Leade with notice naming the leade stating the number of feet claimed describing as near as possible its general course & locality—

Art 8—Any person or person locating any Leade, lode or ledge or deposit of Gold, Silver, Copper, Lead or other metallic bearing quartz rock or clay will be required to file a notice with the Recorder for record within ten days after such location or locations—

Art 9—A monument upon any leade, lode, lodge or deposit will be sufficient to hold an individual or company's claim six months from the date of record—

Art 10—A shaft or Tunnel of ten feet on the leade, lode or ledge of any individual or company's claim will be sufficient to hold the same one year from the date of record & a shaft or tunnel of twenty feet on the leade lode or ledge on any individual claim or company's will be considered a permanent title & not subject to relocation—

Art 11—Any individual or company not worked for one year from the date of record will be declared abandoned & subject to relocation—

Art 12—Any reasonable amount of ground not exceeding ten acres & taken in a body can be located for a mill site for milling & mining purposes together with wood timber & water on the same & must be recorded giving its proper boundaries.

Art 13—No person or company can nor shall in any way obstruct the free access to or from any individual or company claim (mining) nor infringe in any way by cutting wood or timber on ground allowed each side of leade for working & mining purposes—

Art 14—Water claims for milling & mining purposes can be located upon any gulch ravine or spring including & claiming the entire amount of water running in such gulch or ravine to the point of location & amount of water afforded by such spring if actually necessary for milling & mining purposes by the person or company so locating & no person or company can in any way affect or diminish water previously located. Water privileges must be recorded with description of their locality & boundaries of claims—

Art 15—A lot on any town site in this Dist shall be one hundred feet square & subject to preemption & must be recorded—

> Art 16—Grounds for agriculture or garden purposes can be located by filing notice with the Recorder giving the boundaries & location provided such location does not in any matter interfere with the free access to or from any mining claims or interfere with or diminish any water claimed & required for milling & mining purposes—
>
> Art 17—The rights of miners, for milling, mining & working purposes shall have preference over all other rights.[17]

The format of these regulations was frequently repeated during this period, with the exception that the special provisions for millsites, townsites, and agricultural lands may have been unique to the Bradshaw Mountain Mining District.

The requirement for and the content of a notice of location was also frequently addressed, and a typical provision can be found in the regulations of the Fort Rock Mining District, which provided as follows: "All locations made must have a written notice, posted on a monument not less than two feet in height; the notice shall state the number of feet claimed, also the names of the locators and the locality of said claim relative to some natural or artificial object or well established lode."[18]

The most common concern of the districts, however, and probably the area of the widest divergence, was related to the requirements for performance of work on the claims. The most common requirement was that the locator had to perform not less than five days of work on the claim within the first thirty days from and after the date of location, which would hold the claim for six months.[19] Other districts required the excavation of forty cubic feet or a ten-foot shaft to hold the claim for one year.[20] The Bradshaw Mountain Mining District, as indicated above, and the Wallapai Mining District recognized the possibility of obtaining a perpetual title. The Wallapai District's Regulations in this regard provided as follows:

> Art. 9. Each & every claim holder shall be required to put at least two days work upon each location for every two hundred feet claimed within ninety day's after recording—And after the expiration of said ninety days each & every claim holder shall be required to put at least one day's work upon each location for every two hundred feet claimed every thirty days, until the depth of thirty feet is sunk on the location—after the depth of thirty feet is sunk it shall not be subject to re-location, unless publicly abandoned, for one year—[21]

The role of the district recorder continued to be important, and the requirement of the Arizona territorial law that location notices be recorded with the county recorder was occasionally overlooked for expediency's sake. For example, the miners of the Copper Mountain Mining District near Clifton, at their meeting on February 10, 1872, declared: "Resolved, That until a larger force of miners can be introduced here, in order that they may be able to maintain themselves with safety against hostile Indians, the records of this mining district be kept at Silver City, New Mexico."[22]

Finally, it was obvious that the miners not only had clear knowledge of what the right of extralateral pursuit entailed, but also were aware of the legal implications of suggestions for change, because in an amendment to the regulations in the Hualapai Mining District on March 22, 1871, they extended the rights within the district to be "in conformance with Senator Stewart's bill" and provided that *all* veins within the boundaries of the claims that might dip could be followed outside of the sidelines.[23] The fact that this provision did not become law until the General Mining Law was enacted in 1872 was of no consequence.

Post-1872 Mining District Regulations

With the passage of the federal laws, it seems apparent that the miners gained confidence in the new system. In *Rocks to Riches*,[24] mining engineer Charles Dunning and writer Edward Peplow suggested that the mining law of 1872 caused an explosion in staking claims, they suggested an estimated 26,000 claims during the silver boom of the 1870s and shortly after. Dunning and Peplow wrote, "the roaring '70's was the most active period of prospecting and claim filling in Arizona's history." They chronicled the story of eleven of the silver bonanzas from Mohave County's Cerbat Range with Cerbat, Chloride, and Mineral Park; then south to the McCrackin Mine; moving east into central Arizona's Bradshaw Mountain silver producers, the Tiger, Peck, and Tip Top; then further south and east to the biggest single producing mine, the appropriately named Silver King; then into the Globe districts; the Richmond-McMillen Basin; and finally to the biggest silver strike of all, Tombstone. Richard J. Hinton, in his 1877 book, was substantially more reserved and listed fewer than 1,000 mines while suggesting that there may have been 6,500 individual properties that potentially could be called mines.[25]

The most important part of the General Mining Law of 1872, at least in the eyes of the prospectors, was the reaffirmation of their right to enact district regulations: "The miners of each mining district may make regulations not in conflict with the laws of the United States, or with the laws of the State of Territory in which the district is situated, governing the location, manner of recording, amount of work necessary to held possession of a mining claim . . ."[26] The 1872 law did, however, depart from the concept of location of a single vein with a possible total length of 3,000 feet, and fixed the size of a single lode claim at 1,500 feet along the vein and 300 feet on either side of the vein while abandoning the provision allowing associations of lode claimants. Claims were required to be monumented on the ground so that boundaries could be traced, and annual assessment work of not less than $100 annually in order to maintain the validity of the claim was made a federal requirement for the first time. Under the 1866 act, assessment work of $1,000 was required to patent a claim, but no specific requirements were imposed on an ongoing or annual basis. With the 1872 act, however, an annual requirement was imposed, and the assessment work requirement for patent was reduced to $500 per claim. In fact, the mining districts, after the 1872 law, had considerably less legislative authority or discretion other than providing some of the specific details of location of mining claims.

Operating in this restricted area, post-1872 mining district regulations specified whether a shaft was required as a part of the initial location and, if so, the dimensions thereof, how much time would be permitted for the recording of a notice of location, and the value of various types of work for assessment work purposes, as well as providing methods for resolution of disputes and procedural rules for governing the districts.

Given the historical "superiority" felt by the miners within their district, most districts apparently felt that it was necessary to recite in regulations that the laws of the United States, usually specifically reciting chapter 6, title 32 of the Revised Statues of the United States, would be adopted as the laws of the district. In some cases, however, the regulations went beyond the proscription against regulations in violation of the laws of the United States. For example, in the laws of the Silver District in Yuma County, the regulations enacted on January 20, 1879,[27] resolved "that no Chinamen be allowed to work in the District," and the Truman Mining District in Pima County restricted voting within the district to "a citizen of the U.S. or a white citizen."[28] Even as late as 1895, the miners of the Warm Spring Mining District in Coconino County attempted to revert to the 1866 law and restrict the number of claims by enacting a provision "that no Locator shall take more than one claim outside of his claim of Discovery on the same vein; and any other prospector shall locate but one consecutive claim on the same lead or vein."[29] The Pioneer Mining District near Globe, on the other hand, was vehement: "No variation, alteration, or modification of the liberal law of the United States shall be tolerated, its provisions being ample for all purposes."[30]

In taking an overview of the regulations enacted by the miners, it is prophetic that one of the most active "legislative" areas was in the vicinity of the Santa Rita Mountains in southern Arizona, and many of the meetings to enact regulations took place at the Salero Mine, the first mining claim located during the "American" immigration. Within this area, some of the more prominent citizens of the day participated in this process. The meeting at the Salero Mine on February 11, 1875, for example, was attended by H. C. Hodge, Henry Mims, Cornelius Ryan, J. C. Truman, A. C. Benedict, John Mansfield, Joseph King, William G. Boyle, and John E. Magee.

In these new regulations, the recorder of the mining district continued to be the only officer within the district with any official duties. In some districts, however, the recorder's duties were a central part of the location process. A procedure established in the Cedar Valley Mining District in Mohave County at a meeting on September 10, 1874, for example, required the recorder to make a subjective evaluation of the validity of the mining claim and provided as follows: "It shall be the duty of the Recorder of the District to accompany the claimant to the ground located, and before filing it for Record, he shall satisfy himself that it is open for location. Should he think it not open for location, he shall state his reasons for so thinking to the claimant. Should the claimant still insist on having it recorded, the recorder shall make a note of each adverse right, as he thinks exist in regard to the claim, and upon receipt of his fee he shall record it."[31]

Similar types of provisions were enacted the next year in the Smith and Truman Mining Districts in the Santa Rita Mountains where, in addition to determining the lack of conflict, the recorder had to determine whether a discovery existed. The Smith District regulation provided as follows: "Sec. 4: Before recording any mining claim, the recorder shall inform himself by going in person to the claim or lode of every mining location presented for record; and shall examine the claim and see if there is an actual lead or deposit of mineral bearing rock or earth, in the boundaries claimed by the locator, and shall in no case make a record in favor of one party when another party to his knowledge claims the same location."[32]

Independent verification of discovery was also required in the California District, where the location notice had to contain the certification by two witnesses as to the posting, monumentation, and "the discovery of a valuable mineral in a vein lode or deposit."[33]

These types of regulations continued into the 1880s with the Richardson Mining District Regulations requiring the recorder "to be present and superintend the surveying and staking of all lode claims within this district . . ."[34] The Greaterville Mining District also applied additional duties on its recorder by recommending to claim owners that they leave an address with the recorder "and in case the recorder learns of any thing out of order on their claim, he shall notify the owner by postal card."[35]

In addition to participating in the location process, the recorder was also occasionally called upon to inspect the performance of annual assessment work and to provide certificates of such performance. For example, the California Mining District Regulations included the following provision: "When any owner or owners of a lode claim shall have performed the work required by the forgoing section he or they shall prior to the expiration of the time specified in said section, report the same to the District Recorder who shall require such proof of the performance of the work as he may deem sufficient and shall visit and inspect the work upon the claim and if satisfied that the requirements of the law have been complied with he shall thereupon issue a certificate of the performance of the work . . ."[36]

As time progressed, however, most districts eliminated the office of the recorder and designated the local county recorder as the ex-officio recorder of the district. At a meeting of the Wrightson Mining District on March 13, 1881, for example, its regulations were amended with the following provision: "The office of District Recorder should be, and is, abolished . . . , and the recording from this date is to be done only in the office of the County Recorder."[37]

In the few cases where recording was done with the district recorder, it was the local recorder who was responsible for making the filings with the county recorder's office. For example, under the regulations of the Greenlee Gold Mountain Mining District, the district's recorder was required to send a transcript of his records to the county recorder every sixty days.[38]

Most districts established the county records as the official records of the district and provided varying times to record location notices of thirty and sixty days. The miners were frequently concerned with the possibility of "paper filings," and many districts provided that a mining claim was void if a notice was not posted on the claim prior to its being recorded.[39] Recording fees were typically two dollars per claim, although examples can be found for both one dollar and three dollars per claim.[40] Considering that a good suit at the time could be purchased for twenty dollars, these recording fees were of no small consequence.

The requirements for posting and monumenting were reasonably uniform, and the following examples are typical:

> Pajarita Mining District: that all mines hereafter located in this district shall be marked by end monuments or stakes at least 18 inches in height above the ground with sufficient marks placed in or upon them to show to which end of the claim they designate—If stakes are used they must be sunk at least six (6) inches in the ground & have a blaze and figures upon one side.[41]

> Verdi Mining District: That all Ledges or mining claims discovered in this District shall have notices placed on them on some conspicuous place on the mine defining from the point of location in said notice the number of feet claimed and in what direction from the notice said feet are claimed in length, and said notice shall also state the number of feet in width along the vein or lode claimed by the locator and should Locator not mention the number of feet in width, only the minimum number of feet allowed by Act of Congress of May 10th 1872 shall locator be entitled to and that said notice shall be recorded in the Books of the District Mining Recorder within thirty days from the date of discovery of vein or lode.[42]

Some districts had specific work requirements as part of the process of staking a lode claim, although there appears to have been a widespread unwritten custom that a "discovery shaft" was part of this process. This factor can be implied from the occasional requirement that the location notice be posted in the vicinity of the location shaft without making any direct requirement for such a shaft.[43] However, only three mining districts specifically mentioned a requirement for an excavation. In the case of the Silver and Warm Spring Mining Districts, a shaft of five feet was required, while the Cochise Mining District required ten feet. Aside from the implied or specific requirement to dig a shaft, it was not uncommon to require the expenditure of a specific amount for the benefit of the claim during the first three to six months after its location. The language from the Warm

Spring Mining District provided "that every Locator be required to do fifty (50) Dollars worth of work on his claim within six months from date of location; Said work to consist of a hole on the vein or lode of four feet square by five feet in depth."[44]

The value of the performance of location work and assessment work was one of the most frequent subjects of district regulations; the value was fixed at an amount that was normally five dollars or six dollars for each day of work.[45] In some cases, premiums were attached to particular types of work. For example, the Plomosa Mining District in Pima County provided as follows: "The value of assessment work is fixed at $12 per foot in blasting ground and $6 per foot in picking ground."[46]

The provisions for the conduct of meetings for either the conduct of elections or changes to rules and regulations or other matters were almost always covered in the regulations. The most typical requirement for a notice of meeting states as follows: "Ten days prior to the date of holding said [annual] meeting the Recorder shall cause to be posted in three of the most conspicuous places in said District a notice stating the time when and the place where said meeting shall be holden and shall designate in such notice that the meeting will be holden for the purpose of transacting all and every kind of business which may be brought before it."[47]

Special provisions were frequently made, however, if any consideration was going to be given to rule changes, and the Papago District, for example, required thirty days' notice of any rule change. The most frequent requirement restricted any rule changes to annual meetings. For example, the rule in the Pima, Pajarita, and Little Cottonwood Districts provided as follows: "These rules Regulations & By Laws shall not be altered or in any way changed except at a regular annual meeting of the miners of said district, and then only a legal vote of two thirds of all the voters present & voting."[48]

Although a requirement to publish an agenda for meetings was unusual, such a provision can be found in the Pajarita District Regulations, and provides as follows: "A special meeting may be called at any time to transact such business as may be brought before it by posting at least three (3) notices specifying the business for which the meeting is called in conspicuous places in the district signed by the chairman & secty & at least five (5) voters of the District, said notice to be so posted at least ten (10) days before the date of meeting & no business shall be transacted at such meeting other than that specified in the notice."[49]

Very few districts made any requirements for a quorum at meetings. Three exceptions were the Seattle Mining District, which required ten miners for a quorum, and the Greaterville and Globe Mining Districts, which required fifteen miners.[50]

The qualifications of the voters were typically stated as follows: "All persons who are citizens of the US of America or who have declared their intention to become such and own shares of stock or interests in any mines in the district or who has worked in any mine in the district for the twenty days preceding such meeting shall be considered a legal voter and entitled to vote at a miners meeting."[51]

Finally, the mining districts continued the tradition of attempting to resolve any disputes internally and frequently either required disputes to be settled by arbitration or at the very least exhorted their members to use such a process before resorting "to law or force in the settlement of their rights." Some of the typical examples are as follows:

> Smith Mining District: In all cases where disputes or differences shall arise as to the rights of mining claims, the rights of the parties shall first be submitted to an arbitration of seven qualified voters of the district, to be chosen by lot or a less number, if agreed to by the persons who shall sit and hear the evidence and arguments of the respective parties, or their attorneys, and give their decision thereon before the parties shall resort to law or force in the settlement of their rights. And in view of the vexations and ruinous litigation that so often arises over mining claims, every person interested in mining claims in this district, should urge upon each other this mode of settling their disputes.[52]

> Silver Mining District: Resolved: To recommend all disputes arising from locating mining claims, and also all other disputes, to an arbitration of three men in the said district.[53]

> Richardson Mining District: Resolved: That in case of any dispute upon any claim that might arise out of any location made on any lode claim in this Mining District upon complaint being made to the local recorder, he shall have power, and is hereby authorized to appoint two disinterested miners within this Mining District to proceed to an investigation, and with the assistance of these two miners shall have power to decide upon the validity of the adverse claims of the respective parties, and the decision of the recorder in such cases shall be final. In all cases

> of disputed claim titles the recorder's fees shall be $5.00 for each day or fraction of a day so expended in the interests of such parties; such fee to be paid by the party against whom judgment shall be rendered. That all complaints of whatsoever nature or kind, arising out of mining claims by and through any party or parties present, or that may hereafter become interested in mining claims within this Mining District shall be adjusted by the recorder or the recorder and two disinterested miners duly authorized by the said recorder of this Mining District.[54]

Not surprisingly by this time, the mining district regulations continued to address only the location of lode mining claims, and only the Smith and Greaterville Mining Districts made specific provisions for placer claims. This may be significant because the Greaterville District was later formed from portions of the Smith District. The Smith District regulation provided that "all placer or gold bearing gravel claims shall not exceed in extent one thousand feet in length and one hundred fifty in width to one person, and he shall not hold more than one such claim on one gulch by preemption, but shall hold as many as he pleases by purchase."[55] The Greaterville provision also limited placer claims to 1,000 feet by 150 feet "on each side of the gulch."[56] Further, rights to other surface rights were infrequently mentioned. Only the Pajarita and Warm Spring Mining Districts recognized a right to stake millsites, and the Warm Spring regulation allowed the claimant to hold a claim for "timber, water and precious metals within the boundaries of his claim, if mentioned in his Notice of Location."[57]

The "Governmental" Functions of the Mining Districts

With the enactment of the federal mining law in 1866 and, perhaps more significantly, the establishment of territorial government, the mining district likely ceased to be used for any "governmental" purposes apart from regulating the methods of staking mining claims. Whether (or how much) the districts may have or continued to exercise such functions is difficult to determine because of the lack of permanence of any of the district records. Thus, very little evidence of such a function has been found. It is likely, however, that by this time "civilization" had established a sufficient foothold that the miners were content to rely on the local civil government, other than for their desire to decide the outcome of disputes in matters particularly related to the mineral location process. Where, however, the district was either sufficiently isolated or the miners were presumably unhappy with the "law" to be provided by the county government, the governmental function occasionally emerged. One example of this exercise took place at a special meeting of the Greaterville Mining District on January 5, 1893. There, on the southeastern slopes of the Santa Rita Mountains and fairly isolated from the county seat in Tucson, E. N. Fish presided at a meeting whereby Jesus Quiros, after having twice discharged his weapon at a local dance, was determined to be "an undesirable character in this community and that he be ordered to immediately leave this vicinity under penalty of being turned over to the officers of the law for their action. This to take effect within 24 hours from notice." Thus, the ancient penalty of banishment was still alive in southern Arizona.

It was probably unfortunate that the Arizona legislature saw fit in 1866 to repeal chapter 50 of the Howell Code in its entirely. Many of its provisions did not conflict with the federal law and would have lent a great deal of order to the mining law within Arizona—and perhaps provided a model for the other mining states and avoided some of the problems that have lingered in the application of the law. Among the victims of the repeal were the formal requirements for the establishment of mining districts, the resolution of mineral and surface conflicts, and procedures for mining litigation. The result has been that district regulations enacted after 1866 display a wide variation of both content and formality. It also explains why no permanent record of the majority of the mining district regulations exist, as they were probably reduced to writing only in the books maintained by the district recorder and never recorded with the county recorders or published in newspapers.

The "End" of Mining Districts

What remains to be done is to abolish altogether the irregular and whimsical subdivisions known as "mining districts."[58]

With these words in 1879, Rossiter W. Raymond, then the secretary of the American Institute of Mining Engineers, and one of the most respected and influential voices within the early US mining industry, condemned the continued existence of "legislative" authority within the various mining districts recognized under the General Mining Law. Indeed, the existence of authority within the districts was a common criticism of the mining law even in its gestation period. J. Ross Browne, in his 1867 report to the Congress, observed that

Fig. 8. Rossiter W. Raymond (1840–1918). Raymond served as the special commissioner of mining statistics for the Treasury Department during 1868–75. His report in 1869, at 179–223, contained an examination of the history of mining law and gave recommendations in support of the 1866 mining law. In 1867 he became the editor of the *American Journal of Mining* (now the *Engineering and Mining Journal*) and was a national expert on the mining industry. Courtesy of the American Institute of Mining, Metallurgical and Petroleum Engineers

Fig. 9. J. Ross Browne (1821–75). Browne traveled throughout the West from 1860 to 1868. His reports for the Treasury Department contained comments on the mining district laws and the 1866 U.S. mining law. He described Charles Poston's mineral development efforts centered in Tubac in his *Adventures in the Apache Country* (1869). Courtesy of the Arizona Historical Society, #7861.

> in 1853 the Sweetland District was subdivided into three smaller districts, of which North San Juan is one. This latter developed a set of regulations at the time of its organization and adopted the set now in force a year later. A mining recorder was elected in 1854, but he has been absent from the district for five years, and no one has been chosen to fill the place. The regulations are treated by many persons as if they were no longer in force—at least, as regards certain points; and in many cases it would be difficult to ascertain whether there is any good title to claims under the regulations.[59]

The districts in Arizona also had problems. For example, the meeting of the miners in San Francisco District in Mohave County on October 17, 1864, reported:

> Resolved that Messrs. Davis, Thompson and France be appointed a committee to demand the Books from the custodian. The subject was as follows: that the late of election was 28th March 1864 and that the custodian (Reddick) refused to deliver up the Books to any person but Col Akins Depty-Recorder or his order.
>
> On motion it was resolved that the President appoint a committee to demand formally the Books and papers pertaining to the Recorder's Office from the person now holding them. The Pres't appointed the old committee. The committee acted upon and reported as follows:
>
> That the custodian (Reddick) refused "in toto to deliver the Books without an order from the Depty-Recorder, Col Akins. Moved and carried that the meeting be resolved into a committee of the whole to obtain the Books "nolens volens" the same being forcibly and most illegally detained. After some discussion, the above motion was reconsidered.
>
> Resolved that it be right and proper for the Recorder, C. W. C. Rowell procure a new set of Recorder's Books pending the absence of the late Depty-Recorder.
>
> Resolved that the committee be appointed to wait on the ex-Dept'y-Recorder on his return to the District and hand him a copy of the motion on the above meeting and demand from him the Books of Records of the District.[60]

When the issue continued to be addressed in the context of the General Mining Law, Rossiter Raymond, Browne's successor in the office of Commissioner of Mineral Statistics, continued to beat this same drum. In Raymond's 1874 report, he reprinted these rather heartfelt extracts from letter of W. S. Keyes, mining engineer, of Eureka, Nevada:

> Mine-owners in good faith have never, within my knowledge, in a single instance been able to develop a good paying property without being obliged either to buy up countless conflicting and really abandoned claims, or to suffer such harassing and costly litigation as to have rendered even the best of mines a generally ruinous investment. This state of affairs, only too patent to all who have had practical experience in our mining regions, needs here no further elucidation. It may all justly, I think, be attributed to the blundering rules and customs of a generally ignorant body of prospectors, sanctioned and perpetuated by the Congress of the United States....

> Instances are not at all uncommon where individuals or corporations have exhausted themselves financially, and their mine as well, in vexatious lawsuits for the sole benefit of attorneys and hungry witnesses. Too often, unfortunately, the rumor of a legal contest summons from all quarters the birds of prey, whose expectant maw, like the ocean, has no limit of capacity to engulf the golden morsels wrung from the baited victims on either side.[61]

The criticism reached its zenith in a report filed by the Public Lands Commission established by an act of Congress on March 3, 1879.[62] This report, published in 1880, had solicited critical comments, and the continued viability of the district regulations was the subject of several letters.

One of the severest critics was W. H. Beatty, the chief justice of the Nevada Supreme Court and the author of a number of the more significant judicial decisions related to mining claim disputes. Justice Beatty, in a letter to the commission dated November 21, 1879, stated that "the principal, the vital defect in the existing law is the permission to make local laws." He explained:

> I believe that the whole subject of mining locations is an extremely simple one, which may easily and certainly therefore ought to, be regulation by one general law, the terms and existence of which shall be established by public and authentic records, and not left to be proved in every case by the oral testimony of witnesses, or by writing contained in loose papers or memorandum-books, such as are often dignified by the name of "mining records." I am convinced, moreover, that the tainting of every mining title in the land at its very inception with the uncertainty which results from the actual or possible existence of rules affecting its validity, perfectly authentic evidence of which is nowhere to be found, is a stupendous evil.[63]

Justice Beatty also pointed out that even where regulations were in writing, questions remained whether they had been regularly adopted or generally recognized by the miners within the district. He further noted:

> There may be two rival codes, each claiming authority and each supported by numerous adherents; evidence may be offered of the repeal or alteration of rules, and this may be rebutted by evidence that the meeting which undertook to effect the repeal was irregularly convened or was secretly conducted in some out-of-the-way corner, or was controlled by unqualified persons; customs of universal acceptance may

> be proved which are at variance with the written rules; the boundaries of districts may conflict, and within the lines of conflict it may be impossible to determine which of two codes of rules is in force; there may be an attempt to create a new district within the limits of an old one; a district may be deserted for a time and its records lost or destroyed; and then a new set of locators may reorganize it and relocate the claims.[64]

Justice Beatty's fears were not academic, and an example of the "two district" conflict can be found in the public records of Pima County, when the Tucson Mining District was formed at a meeting held at Camp Sacramento on April 30, 1878, including substantially the same ground as the Harshaw Mining District established at a meeting held a day earlier at the French millsite. Apparently, this overlap was discovered, as the Tucson District appears not to have subsequently enacted any regulations.[65]

The calls to the Public Lands Commission for abolition of mining districts as a rule-making authority and records repository were almost unanimous.[66] Some of the testimony presented to the commission provides a good insight as to how the mining district "really" worked. Some of the typical problems were detailed by B. C. Whitman of Virginia City, who, testifying in San Francisco, on October 11, 1879, explained:

> All these notices have been transcribed, and the originals have been held in the office of the county recorders. These were originally drawn by the claimants themselves, with a great indefiniteness of description and date. You would see "recorded this day," and there would be no date. It was posted on the claim and filed with the local mining recorder. There was no security for the integrity of the records in the hands of the mining recorder. Those records will show the very great carelessness with which the business was done. For instance, in the notice of the Miller grant, as it is called, at least one-third of the names that were originally written upon the notice have been scratched out and other names substituted; properly enough, because the names of most of his friends were written there, and as those friends did not come on time he would scratch out their names and insert others. He used the names of his friends to take up a number of additional feet.[67]

Alfred James, a land office register, testifying in Los Angeles, on October 17, 1879, also recommended abolition of the districts:

> The mining records under the local regulations are copied in a very loose and unsatisfactory manner. Mining recorders are generally irresponsible persons, not under bonds, and are easily used for the purpose by those that wish to evade the law. Mining districts are irregularly organized. There is no established rule as to the number of persons necessary to organize a district. It may be by three persons, and I have known districts organized by two miners at a miners' meeting of themselves—the meeting called by themselves upon notice sent to each other; one presides and the other is secretary. There is no evidence needed as to whether they are miners or whether they are citizens; all this has to be assumed.
>
> Having elected themselves president and secretary they pass a code of laws to govern the district, and pass such laws as will subserve their own interest, and proceed to record the claims they cannot hold themselves by "dummies."
>
> It often occurs that a mining recorder is elected, makes a few records, abandons the district and takes the records with him.
>
> The boundaries are designated without survey, often by illiterate men, and so very frequently overlap boundaries of other districts. It also occurs that a second mining district is organized, and the old laws repealed and new ones enacted.[68]

A single halfhearted dissenting voice heard was that of Aaron A. Sargent, a former representative from California, a lawyer, and a miner. He had introduced both the 1870 and 1872 mining laws,[69] and he testified that "great care should be taken in abolishing 'miners rules and regulations.' They contain a good deal of practical wisdom. The present law embodies some of them, of general application, and perhaps still more could be utilized. I should recommend this rather than their abolition."[70]

Rossiter Raymond was again in the middle of the criticism and clearly in the camp of the majority. Ten years earlier, as commissioner of mineral statistics, he had some definite opinions that he had expressed to the Congress:

> What are these mining customs to which the law pays such sweeping respect? They are edicts passed at 24 hours notice by mass meetings of from 5 to 500 men; it requires no more formalities to abolish or amend them than it did to make them—a notice pasted on a door, a "mass meeting" next day, and the thing is done. The records of titles

> are kept by an officer called the recorder, not known to the law nor answerable for malfeasance in office, except that if he were known to tamper with the books and is charged his life might be taken by the party wronged. The records are kept in a few districts in fireproof offices and in suitable form, but more frequently in small blank-books, pocketbooks or scraps of paper, stowed away under the counter or behind the flour-barrel or the stove of a store or bar-room. These are not exaggerations. The title to millions of dollars worth of property depends on records no better cared for than this.[71]

In Raymond's open letter transmitted to the Public Lands Commission, he referred to the mining district regulations as one of "a class of evils inherited from . . . local customs." He viewed as absurd the situation "of permitting the title to mineral lands to rest upon the shifting and untrustworthy basis of an irregular, periodical plebiscite, the edicts of which are carried out by irresponsible officials, and the records of which often may be, and often have been, exposed, without efficient guardianship, to loss, destruction, mutilation, or falsification." He concluded: "What remains to be done is to abolish altogether the irregular and whimsical subdivisions known as 'mining districts,' with all their officers, and make all mining titles on the public lands originate in entries duly attested and preserved in duplicate or triplicate by the regular officers of the United States."[72]

Raymond's final suggestion was that it was the business of local governments to regulate mining operations in lieu of the United States establishing a comprehensive mining code.

Statewide Mining Laws Supersede the District Regulations

The net action taken related to the mining districts as a result of the commission's report was nil. The final report had recommended that districts be abolished and that location notices be filed with the United States surveyor general for the surveying district in which the claim was situated.[73] As it turned out, however, the critical comment apparently generated or encouraged interest by the various state legislatures to follow Raymond's suggestion and provide the implementation of the federal law on a statewide basis that the district regulations had previously provided piecemeal. These acts, beginning with a model enacted in Colorado in 1874,[74] established a pattern for a statewide implementation of the federal mining laws.

The seeds for a uniform law in Arizona had been planted in 1867 with a call for a "Miner's Convention" published in the *Arizona Miner*.[75] The suggestion was that the legislature assemble "a Convention of miners with power to frame a code of quartz laws for their respective counties." The *Miner* argued that "the way our mines are managed at present and the manner in which laws have been and will continue to be made, unless the Legislature sanction the proposed measure, is calculated to breed much litigation and cause a want of confidence and feeling of insecurity among miners and mill-men. . . . Give us one law for each county, so that owners may go to bed and wake up next morning without fear that the law under which they, in good faith, located and worked their claims will not be changed whenever it suits the purpose of persons advance to their interests." On March 20, 1895, the *Miner*'s seeds bore fruit when the Arizona Territorial Legislature enacted legislation to provide uniform mechanics for the location of mining claims.[76] The absence of any prior action by the Arizona Territorial Legislature may have been recognition of needed autonomy for the various mining camps. The boom in gold, silver and copper mining between 1867 and 1892 in Globe, Harqua Hala, Jerome, Ray, Silver King, Tombstone, and Vulture was, for the most part, within existing districts, and miners would have likely viewed new laws of general application as undermining rights that had been established under the old laws.[77] Also, during the 1870s, the format for the creation of mining district regulations was widely circulated.[78]

The 1895 Arizona law departed from the concept of retention of the district regulations and followed the Colorado example and similar laws that had been enacted in New Mexico in 1875, South Dakota in 1877, and Wyoming in 1886.[79] The law consisted of thirteen sections, and provided instructions for the contents of a location notice; required the locator, prior to recording the notice to sink a discovery shaft on the claim, "to a depth of at least ten feet from the lowest part of the rim of such shaft at the surface, and deeper if necessary, until there is shown by such work a lode deposit or mineral in place"; and to delineate the boundaries of the claim with substantial monuments. All of this work was to be completed within ninety days from and after the date of "discovering the lode and the posting of the notice thereon." The law also allowed for the amendment of certificates to correct errors and to take in additional ground; provided a form for a permissive filing of annual assessment work in the county records, which, if filed, would constitute "prima facie evidence of the performance of such labor or the making of such improvements"; and finally established a procedure for the relocation of forfeited or abandoned claims, providing essentially for the deepening of the discovery shaft an additional ten feet and the reestablishment of the boundaries of the claim, as well as posting a new location notice stating "if the whole or any part of the new location is located as abandoned property."

The 1895 law also followed the tradition of ignoring placer deposits, but two legislative sessions later, on March 2, 1899, separate provisions established a procedure for the location of placer claims.[80] The claims were required to be monumented at each corner, and although the law did not specifically state this, it implied that locations should be made according to the public land survey system, because the number of acres within the location should be stated, and allowed a procedure for witness monuments were the true corners were inaccessible. Also, unlike the ninety days required for lode claims, the placer law required that the location notice be recorded within sixty days from the date of location. This seems logical because no location work was required.

A precursor to the 1895 legislative effort needs to be recognized and may provide an insight into the resolution of a number of issues facing mining claimants. This effort was the Superstition Mining District, whose organizational meeting minutes were recorded on June 8, 1894.[81] In following the traditional measurement of claims, the regulations also recognized the potential of making discoveries through an adit so long as it was more than four feet underground, recognized a right-of-way for pipelines and tramways, and made clear that abandoned adjacent ground required the recording of an additional location certificate.[82]

This legislative effort was apparently accepted by the mining districts, because, subsequent to its enactment, no new substantive regulations were enacted by any of the mining districts—and only an occasional boundary description was recorded in the public records. The sole exception was the Warm Spring District, which enacted bylaws on April 16, 1896, but this event can probably be attributed to an unawareness of new legislation.[83]

During and after the development of this legislation, the mining districts continued to multiply, but only in the minds of the miners, and toward the end of the nineteenth century, although a number of new mining districts continued to appear on maps, most were either never formalized or only established insofar as their boundaries were placed of record. Therefore, as of 1912, when a comprehensive mining district evaluation was performed by the United States Geological Survey, 137 districts were shown in Arizona, and this number had increased to 246 when a similar review was undertaken by the Arizona Bureau of Mines in 1961.[84]

General Legislation Regulating the Mining Industry

Even during the 1866–95 era, while the authority of the mining districts to dictate the steps required to establish mining claims on the public domain of the United States was unchallenged, the Arizona territorial legislature did not feel precluded from enacting laws of general application. These territorial laws could be best characterized as laws that governed the relationship of mineral claimants as between themselves, as opposed to establishing rules and regulations for the basic initiation of rights on the public domain.

The enactment of these territorial laws began in the very first session of the legislature where the rights of persons in the military service were protected against any mining districts who might seek to disqualify members in the military service. Thus, the legislature declared that "the laws of any mining district contrary to the spirit and provisions of this act are declared to be null and void."[85] In 1867, provisions were made permitting co-owners of lode claims, upon thirty days' notice to any co-owner who was not contributing to the working or development of the property, to petition the court for a segregation or partition of the property. Thus, after the expiration of sixty days, if a notified party refused to join in the work, the court was required to appoint two commissioners (if the commissioners did not agree, they would appoint a third) to "segregate" the claims of the parties refusing to join in development, and the court would thereafter enter a decree based on these findings.[86] The law obviously had little support because its application was limited to Mohave County.

Mining property was taxed under Howell Code as any other property, provided, however, that individuals could elect to pay a tax applied to corporations of 5 percent of the net proceeds tax plus fifty cents for each hundred dollars in value of real estate in lieu of all other ad valorem taxes.[87] The law was repealed two years later and replaced in 1875 by another tax on net proceeds from the mines.[88] Basically, this act provided that any ores, tailings, or mineral-bearing materials of whatever character were assessed for taxation based upon deductions from the gross yield. The first deduction was of the "actual cost" of extracting the minerals from the mines, the actual cost of "saving" the tailings, the actual cost of transportation to the place of reduction, and the actual cost of reduction of sale, with the remainder deemed the net proceeds. The act provided, however, that the total amount of deductions could not exceed the gross yield, and on materials with a return of more than $30 and less than $60 per ton, the deductions could not exceed 90 percent of the gross yield. On materials with a value of more than $60 and less than $100 per ton, the deductions could not exceed 80 percent of the gross yield; on materials of a value of more than $100 but less than $200, the amount could not exceed 60 percent; and on all materials with a value of more than $200 per ton, the total amount of deductions could not exceed 40 percent of the gross yield. An exemption of $20 per ton was allowed for all ores, tailings, or minerals that were roasted before reduction. Forms were provided to the miners and mining companies that had to be returned under oath specifying the costs and gross returns, which were required quarterly. If the statement was not made, it was the assessor's duty to provide an estimate from the best sources within his reach as to the amount of minerals extracted and assess the same, an assessment that was binding. The tax that was levied for territorial and county purposes was two dollars on each hundred dollars of net proceeds. The apportionment of this tax was amended on February 9, 1877, to provide that 25 percent of the amount would be apportioned to the territorial treasury for general purposes, and 75 percent would be apportioned to the county treasury for a road fund to be used and disbursed by the Board of Supervisors for construction and repair of public roads, highways, and bridges.[89]

In 1881, locators of mining claims were granted "the right of way over all adjoining or adjacent mines or mining-claims for the purpose of transporting supplies, material, or ores used upon or taken from the claim or claims so entitled to the right of way; and it shall be lawful in the exercise of this right of way to construct such a road, tramway, or railway as may be necessary to transport such supplies, materials, or ores."[90] The owner of the adjoining claim was however, entitled to remuneration as provided under separate toll road legislation.

Another law passed in 1881 created the office of a "territorial geologist" to collect information on the condition of the mining industry, including character and efficiency of beneficiation and metallurgical processes, to collect and exhibit a cabinet to represent the ores and minerals of the territory.[91] The geologist was also required to assay ores and charge ordinary business rates for such services, as well as furnish the legislature with information and statistics.

Other laws related to the protection of mining property included:

- Destruction of notices of location was a criminal offense and punished as a misdemeanor.[92]
- Water in mines could be a significant problem, and where common subterranean communication of water existed, the law provided that "it shall be the duty of the owners, lessees or occupants . . . to provide for their proportionate share of such drainage, or to prevent the water in such mine from flowing in or upon neighboring mines, thereby imposing upon them an unjust burden."[93] This same law allowed the right to maintain an action to recover damages and notice of joint operations, but did not apply to unopened or undeveloped mines.
- In 1895, "salting ores" was made a criminal offense punishable by fines of not less than $500 nor more than $1,000 and imprisonment of not less than one nor more than ten years.[94]
- A procedure for optional sales of mining property belonging to estates or guardianships was established whereby the optionee could be assured of obtaining a deed after satisfaction of obligations under a sales agreement.[95]

The Evolution of the Claim-Staking Process

Since the 1895 and 1899 laws, the process of actually staking a mining claim on the public domain of the United States has changed very little. The only real changes have related to the pit originally required to be sunk as a part of the location process for a lode claim. This pit, originally required to be a depth of 10 feet, was rather inexplicably reduced to 8 feet in 1909, while at the same time recognizing that "any open cut, adit or tunnel" of 192 cubic feet would comply with the statute.[96] These 8-foot location pits thereafter dotted the public domain in mineralized areas. Where lands had been patented under the provisions of the Stockraising Homestead Act of 1916 or otherwise exchanged where the mineral estate remained subject to entry under the federal mining law,[97] these exploration pits constituted an ongoing source of real or hypothetical dissension between miners and the livestock industry.

In 1964, in recognition of the fact that these pits rarely served as an indication of the presence of a mineral discovery, prospectors were given an alternative of drilling a 10-foot hole on each claim in lieu of digging an

8-foot shaft. This work could also be aggregated for up to ten contiguous claims (so long as they could be included within a 3,000-foot block) where a 100-foot drill hole could be substituted.[98] This alternative was not entirely acceptable, because the process of drilling 100-foot holes for claim blocks still did not provide the exploration industry with much meaningful information as most mineral horizons for the low-grade copper deposits then being sought were substantially below 100 feet. Further, the construction of roads required to get truck-mounted drilling equipment into all portions of the claim block being staked continued to create the same unnecessary conflict between mineral claimants, federal regulators, and surface owners. Therefore, beginning in 1975, a group of mineral explorationists based primarily in Tucson, as members of the Southwestern Minerals Exploration Association (SMEA), in alliance with a number of environmental groups, prepared amendments to the mineral location law and began lobbying for the elimination of all location work and to substitute the preparation and recording of a map of the area claimed.[99] These efforts were finally successful in 1978, and all location work was eliminated, with a map required as a part of the location process.[100] This effort was probably helped substantially by the fact that a map filing requirement had also been mandated by federal legislation in 1976, and it was a source of satisfaction to the backers of the Arizona legislation that the language of the federal regulations implementing the 1976 federal law was influenced by the language of the Arizona bill.[101]

This new law also allowed the recording of a second map filing whereby the owners of existing claims, by recording a map before October 21, 1980, would be entitled to "a rebuttable presumption that the claim was monumented on the ground so that its boundaries could be readily traced."[102]

This 1978 law made a number of other changes. These changes included:

- moving the location monument from the center line of lode claims to one corner designated in the location notice;
- requiring the recorded location notice to be signed by the locator;
- defining what constituted a "substantial monument" for purposes of the location and boundary markers.

The idea of the corner posting of location notices was that this encouraged the creation of "clusters" of location notices because most claims were located as part of large blocks. Thus, location notices could be more readily found. The basis for the actual signature on the recorded notice was premised on the language of the prior law that required the locator to record "a copy" of the notice posted on the ground with the county recorder. Thus, a carbon or photocopy of the original was technically capable of being recorded in the official county records. By requiring an actual signature, the integrity of the recording process was felt to be enhanced.[103]

The definition of the substantial monument was intended to clear up arguments within the industry as to what was "substantial" and to legitimize existing practices. The problem was that the placer claim statutes required the use of four-by-four posts, and many felt that anything less was illegal. Anyone who carried these posts as part of the location process was therefore eager for some lighter alternative. Thus, permitted monuments included material readily distinguishable as a monument and not less than one and a half inches in cross section. This dimension was designed to legalize the use of two-by-two posts and plastic pipe. The law was also amended to require, as had been the custom in many of the mining districts one hundred years previously, that each monument was required to contain an identification "of which corner or end center of the claim or claims for which it was erected." Unfortunately, in the process of drafting the legislative bill, which contained strikeouts and additions to existing laws, a reference to lode claims requiring "six monuments" was left in the language, and the final version of the 1978 law required six boundary monuments for all mining claims. This was fine for lode claims with four corner monuments and two "end center" monuments defining the intersection of the centerline of the claim with the "end line"—but was not required as a part of the location process of placer claims and millsites, which only required corner monuments. Thus, although the statute required the use of six posts, instructions were only given for the placement of four in the case of placer claims and millsites.

A problem regarding the corner-posting requirement was eventually raised by Harvey Smith, a US deputy mineral surveyor from Phoenix, who argued that the Bureau of Land Management was requiring a centerline posting before any lode claim could be submitted for patent. The regulation was arguably only illustrative and clearly contained other extraneous requirements (including a location shaft) that were being universally ignored. The fact of the matter was, however, that the corner posting was an administrative problem and therefore, in 1989, the legislature was persuaded, primarily through the efforts of Smith, who was also then a member of the Arizona Department of Mineral Resources, to move the location notice for lode claims back to the centerline.[104]

Finally, this most recent occasion for amendment was used to solve the mystery of what to do with the two unknown posts for placer claims and millsites and, while saving a corner posting for such claim, specified that "only the corners or angle points of the claims must be monumented."[105]

The process of staking a mining claim has always been an effort of the miners to insure the existence of evidence of prior possession and a good-faith intention to locate and work a claim. The speculator was spurned and miners were frequently admonished to follow the rules. In its January 26, 1867, edition, for example, the *Arizona Miner* preached from an Idaho report that "indefiniteness and non-compliance with the local law in locating quartz claims have led, and will continue to lead to vexatious lawsuits; also, retard prospecting."[106] The same newspaper, later editorialized that "it would be well for claim owners to see that their claims are properly staked and that the boundaries are well defined . . . If it is wrong, the locator will be the loser. If the ground be carefully surveyed before the record is made, no trouble will be likely to arise from getting the boundaries entangled with adjoining claimants . . ."[107]

The responsibility for the promulgation of these rules has been a matter of individual expression, and perhaps the absence of governmental infrastructure into the widely scattered mining areas was reason enough to disregard the benefits of uniformity. By the early 1890s, there is evidence that approximately one hundred mining districts had either been formally organized or had functioned officially since the formation of the first districts in the "Arizona" portion of New Mexico Territory in 1862 scattered throughout the territory but generally grouped into the regions of western Arizona (essentially the mines along the Colorado River), southern Arizona (most of which were in either the Santa Cruz and San Pedro River valleys or adjacent areas), and central Arizona (the Prescott and Globe areas). The sheer number of districts and improving governmental organization seems to have clearly paved the way for the uniform laws enacted in 1895 and 1899.

The uniform laws did not spell the death of the districts, however, as the "whimsical" mining districts remained as a nostalgic geographic description that is almost universally used even in modern times as a part of the "legal" description of mining claims and by technical writers to identify a mineralized area.[108]

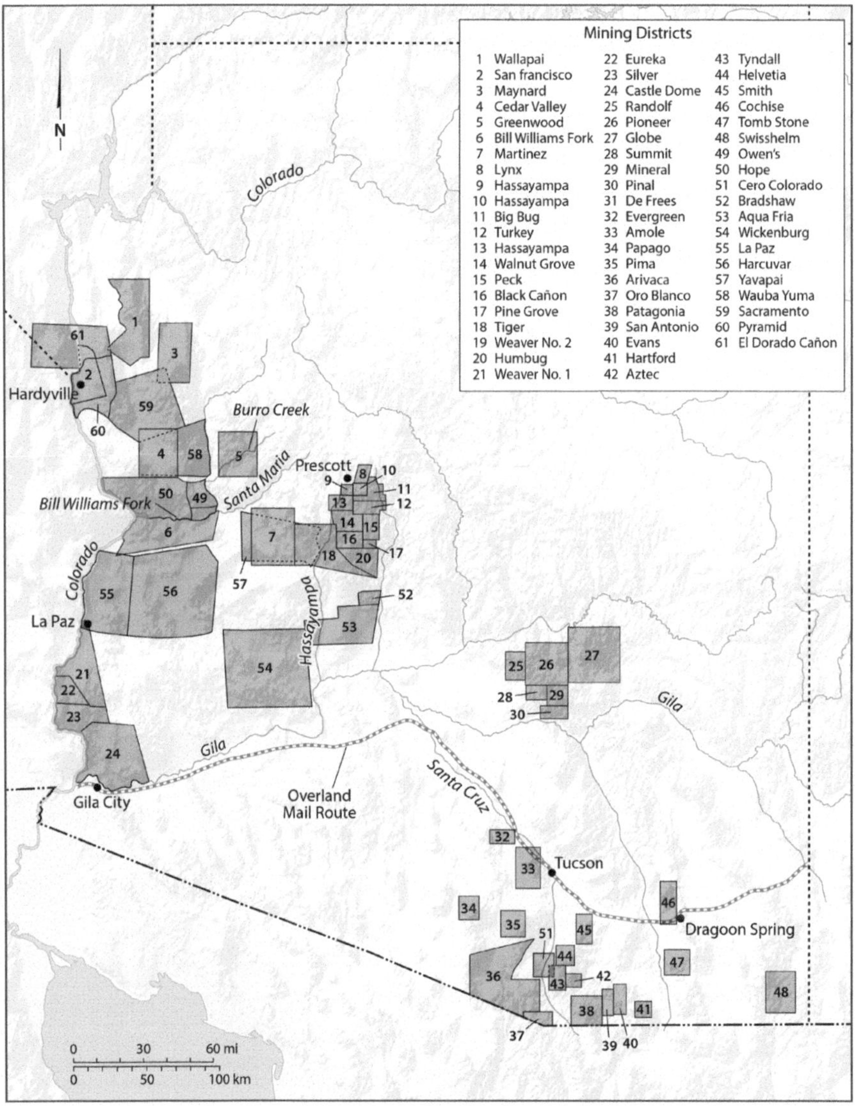

Map 1. Mining district boundaries. Map by William Nelson.

APPENDIX A

Mining Districts of Arizona

Appendix A is an alphabetical listing of the mining districts in Arizona that were either formally organized or otherwise recognized by the miners during 1862–1912. References to 182 separate districts have been found, but approximately only the full text of 70 exist. I have provided the full text of extant rules and regulations in appendix B. The reason for the chronological listing in this Appendix A is to allow the reader to easily follow the efforts of individuals who were active in multiple mining districts during a specific time frame.

The heading for each of the districts in this appendix includes the date of organization, if known, and reference to the sources for the existence or maps indicating the location of the district and other names of the district. The boundary description are either contained verbatim from regulations themselves or are as noted on various maps. Many of the latter-day districts are simply place names, and the district was never formally organized. The location of the district by reference to the county is according to the county boundaries of the State or Territory of Arizona as existed when the district was organized. The date of the establishment of the district is based either on a date of organization from the district regulations, but "pre-" indicates that the district is shown on a map of that date or a location notice was found of record on that date using the district's name. The sources of information used for the compilation of this appendix were either from newspapers (where the date, page, and column is cited), the official records of the various counties (cited to the book and page of recording), documents in the archives of the Arizona Historical Society, Mohave and Yavapai County archives, the records of the Bureau of Land Management or the General Land Office, or materials prepared by Clarence King and published as a part of the Tenth (1880) Census of the United States.

The key to the references of sources used below are:

1863 GLO INDEX: Index to Arizona territory mining district regulations existing in 1863 as found as the first document in General Land Office (GLO) Microfilm.

1866 GLO REPORT (1866 GLO MAP): "Sketch of Public Surveys in Arizona and New Mexico to Accompany the Report of the Commissioner of the General Land Office, 1866," General Land Office, October 2, 1866.

ABM MAP: E. D. Wilson, R. T. O'Haire, and F. J. McCrory (compilers), "Map and Index of Arizona Mining Districts" (The University of Arizona, Arizona Bureau of Mines, August 1961).

BLM RECORDS: Records of the Arizona State Office, Bureau of Land Management, Phoenix, Arizona.

GLO MICROFILM: Carton #22398, General Land Office microfilm records on deposit with Arizona State Archives, Phoenix.

HINTON: Richard J. Hinton, *Handbook to Arizona 1877* (Rio Grande Press. Glorieta, NM, 1970, rep. of 1878 ed.), map prepared by Dewey & Co. and published in the *Mining and Scientific Press.* Hinton also included separate maps within the text noted hereafter by page number.

KING: King, Clarence, *The United States Mining Laws and Regulations Thereunder, and State and Federal Mining Laws to Which are Appended Local Mining Rules and Regulations* (10th U.S. Census Report, GPO, 1885).

RAND, MCNALLY: "Map of Arizona," Rand, McNally & Co., Engrs., Chicago, 1881.

USGS REPORT: James M. Hill, "The Mining Districts of the Western United States," USGS *Bulletin* 507 (Washington, DC: GPO, 1912), 54–76.

ARIZONA MINING DISTRICTS

Ajo, est. ?, Pima County, shown on USGS Report, ABM Map, and Hinton.

Amole (Sierra Amole), est. pre-1872, Pima County, generally includes Tucson Mountains. *See* Sierra Tucson, Tucson Mountains; boundaries shown on Hinton.

Agua Fria (Agua Frio), est. 01-11-1865, Yavapai County.

Boundaries: From Regulations: The boundaries of this District shall be as follows—On the North by Bradshaw, Turkey Creek and Walnut Grove Districts. Thence running southward along the eastern line of the Wickenburg District fourty miles, Thence east ward fourty miles crossing the Agua Frio at or near the Frog Tanks, thence north fourty miles, Thence westward, to the south east corner of the Bradshaw district.

Agua Fria, est. ?, Maricopa County, shown on USGS Report, ABM Map.

Aravaipa, est. ?, Graham County, shown on USGS Report.

Arivaca, est. 04-29-1877, Santa Cruz County.

Boundaries: From Regulations: Commencing at a monumented stake surrounded by stones near the buildings of the Arivaca Ranch and running thence 10' West of South to the Sonora line, again commencing at the monumented stake and running thence East 10 miles which is the East Corner, thence North 15 miles which is the North East Corner, thence North West to the intersection of the Main road from Tucson to El Plomo, which is the North West Corner, and thence along said road to the Sonora line. *See* Hinton.

Ash Peak (Duncan), est. ?, Graham County, shown on USGS Report, ABM Map.

Aubry, est. pre-1873, Mohave County, shown on Rand, McNally.

Aztec, est. 06-09-1877, Santa Cruz County.

Boundaries: From Regulations: Commencing at the Easterly end of the Empress of India Mine and running west of north along the Eastern boundary of the Tyndall Mining District to the Eastern end of the Georgia Mine in the Tyndall Mining District Santa Rita Mountains, thence due North two miles, thence due East three miles, thence South six miles, thence Westerly to the point of starting at the Eastern end of the Empress of India Mine in the Tyndall Mining District Santa Rita Mountains. *See* Hinton 127.

Baboquivari, est. ?, Pima County, shown on USGS Report, ABM Map.

Banner (Troy, Christmas), est. ?, Gila County, shown on USGS Report, ABM Map.

Beaver, est. 06-13-1866, Yavapai County.

Boundaries: From Regulations: Commencing on the Agua Frio river, at the mouth of Coyote creek and following up the Agua Frio, north, to the mouth of the Big Bug creek; thence west to the head of Rabbit creek; thence south to the head of Coyote creek; thence east, down said creek, to the Agua Frio.

Regulations published in *Arizona Miner*, 5:3 (June 13, 1866).

Bentley, est. pre-1873, Pima County.

Bill Williams Fork, est. ?

Boundaries: *See* Hinton.

Regulations: Unknown.

Big Bug, est. 02-07-1865, Yavapai County.

Boundaries not stated but seemingly a redesignation of the Woolsey Mining District. *See* Hinton.

Black Cañon, est. pre-1870, Yavapai County, shown on Rand, McNally, USGS Report, and ABM Map, referred to as being adjacent to the east of the Bradshaw Mountain District in regulations of the Bradshaw Mountain Mining District enacted October 24, 1870. *See* Hinton.

Black Hills, est. ?, Yavapai County, shown on USGS Report, ABM Map.

Black Rock, est. ?, Yavapai County, shown on USGS Report, ABM Map.

Black Warrior, est. ?, Gila County, shown on USGS Report, ABM Map.

Bloodsucker, est. pre-1880, Pima County.

Blue Mountain, est. pre-1871, Yavapai County.

Regulations: Unknown.

Referred to in *Arizona Weekly Miner*, 2:3 (Mar. 18, 1871).

Bradshaw, est. 09-14-1864, Yavapai County.

Boundaries: From Regulations: The district was declared to be ten miles square, commencing at a monument in the town of Montezuma, on Turkey Creek, which shall be the centre of the district; and this in open meeting was declared the limits of the boundaries of the said mining district of Bradshaw.

Bradshaw Mountain, est. 10-24-1870, Yavapai County, shown on USGS Report.

Boundaries: From Regulations: Bounded on the north by Turkey Creek Dist, on the west by Bradshaw Dist on the south by Pine Grove Dist & on the east by Black Cañon Dis't.

Regulations published in King, 267.

Bunker Hill (Copper Creek), est. pre-1880, Pinal County, shown on USGS Report.

Burro, est. pre-1876, Mohave County.

Cababi (Comobabi), est. ?, Pima County, shown on USGS Report, ABM Map.

Boundaries: Unknown but recited as being sixty miles west of Tucson.

California (Paradise, Chiricahua), est. pre-1864, Cochise County.

Cañon del Oro, est. 08-03-1880, Pinal County.

Boundaries: Unknown, but center of district is in Section 23, Township 10 South, Range 14 East

Regulations shown on Pima County index records as being recorded on above date but cannot be found in records.

Casa Grande (Vekol), est. pre-1881, Pinal County, shown on USGS Report, ABM Map.

Castle Dome, est. 12-08-1862, Yuma County.

Boundaries: From Regulations: That the District wherein said veins are situated be called the Castle Dome District and be bounded as follows: Beginning at the Peak known as Castle Dom—thence 10 miles south—Thence East 10 miles—Thence north 10 miles—Thence West 10 miles to the starting point. *See* Hinton, 155.

Cave Creek, est. ?, Maricopa County, shown on USGS Report, ABM Map.

Cedar Valley, est. pre-1864, Mohave County.

Boundaries: *See* Hinton.

Cerro Colorado, est. 04-23-1864, Santa Cruz County, shown on ABM Map.

Boundaries: From Regulations: To extend 12 miles east, 12 miles west, 12 miles north, and 12 miles south, from said Cerro Colorado mine, and shall be 24 miles square.

Regulations published *Weekly Arizona Miner*, 1:2, 3 (Aug. 24, 1864).

Cherry Creek, est. pre-1881, Yavapai County, shown on Rand, McNally, USGS Report, and ABM Map.
Boundaries: *See* ABM Map.
Regulations: Unknown.

Chimehueves (Chemehuevis), est. pre-1878, Mohave County, shown on USGS Report, ABM Map.

Chiricahua, est. pre-1878, Cochise County, referred to in regulations of the California Mining District enacted September 7, 1878.

Cienega, est. ?, La Paz County, shown on USGS Report, ABM Map.

Clark, est. ?, Graham/Greenlee County, shown on USGS Report, ABM Map.

Clifton, est. pre-1881, Greenlee County, shown on Rand, McNally.

Cochise (Johnson), est. 12-26-1878, Cochise County, shown on USGS Report, ABM Map, and Hinton.
Boundaries: From Regulations: Starting on east side of Small Pass about 5 miles East of Rocky Spring or Fink Mill Site and running thence Northerly along east of Dragoon Mountain about 12 miles to Pass Alamos and Silver City Road. Thence westerly along same road 6 miles. Thence South about 12 miles to Butterfields Road. Thence Easterly about 6 miles to point of commencement. *See* Hinton.
Regulations recorded in Book 1 (Misc.), page 604, Pima County Records.
Other names: Johnson.

Colorado, est. 01-08-1863, Mohave County.
Boundaries: From Regulations: Commencing at the Mouth of Eldorado Cañon running Ten miles North following the Course of the river, thence West Twenty miles thence South Twenty miles thence East Twenty miles to the River, thence North Ten miles following the Course of the River to the place of beginning.
Regulations on deposit with Mohave County Historical Society.

Congress, est. ?, Yavapai County, shown on USGS Report, ABM Map.
Boundaries: Unknown.
Regulations: Unknown.
Other names: Martinez.

Copper, est. 06-20-1882, Maricopa County.
Boundaries: From Regulations: Commencing at a point when the Section line between Section 14 and 15 in T.1 North of R.4 East crosses the main channel of Salt River thence south on said line to the south line of Maricopa County; thence west on said Maricopa County line to the main channel of the Gila River; thence down said Gila River to the junction with Salt River; thence up the main channel of said Salt River to the place of beginning.
Regulations recorded in Book 1 (Misc.), pages 349–53, Maricopa County Records.

Copper Basin, est. ?, Yavapai County, shown on USGS Report, ABM Map.
Boundaries: Unknown.
Regulations: Unknown.

Copper Mountain, est. pre-1864, Greenlee County, shown on 1863 GLO Index, USGS Report, and ABM Map.
Boundaries: From Regulations: From a point on the Frisco River known as Lezinskys Dam in a direct line Westerly to the stone cabin on Chases Creek thence Northerly following the bed of said Chases Creek, to its source, thence continuing Northerly to Walnut Spring.
Regulations dated July 20, 1872, on GLO Microfilm and recorded in Book 1 (Misc.), page 1, 56–57, Graham County Records.
Other names: Morenci.

Courtland, est. ?, Cochise County, shown on USGS Report, ABM Map.
Boundaries: Unknown.
Regulations: Unknown.
Other names: Turquoise.

DeFrees, est. pre-1881, Pinal County, shown on Rand, McNally, Hinton.
Boundaries: *See* Hinton.
Regulations: Unknown.

Del Rio, est. ?, Yavapai County, shown on USGS Report, ABM Map.
Boundaries: Unknown.
Regulations: Unknown.
Other names: Granite Creek.

Dos Cabezas, est. pre-1881, Cochise County.
Shown on Rand, McNally, USGS Report, and ABM Map.
Boundaries: *See* Rand, McNally map.
Regulations: Unknown.

Dragoon, est. ?, Cochise County, shown on USGS Report, ABM Map.
Boundaries: Unknown.
Regulations: Unknown.
Other names: Golden Rule.

Dripping Springs, est. ?, Gila County, shown on USGS Report, ABM Map.
Boundaries: Unknown.
Regulations: Unknown.

El Dorado Cañon, est. pre-1866, Mohave County, shown on 1866 GLO Map, USGS Report, and ABM Map.
Boundaries: *See* 1866 GLO Map.
Regulations: Unknown.
Other names: Eldorado Pass.

Ellison, est. ?, Gila County, shown on USGS Report, ABM Map.
Boundaries: Unknown.
Regulations: Unknown.

Empire, est. ?, Pima County, shown on USGS Report, ABM Map.
Boundaries: Unknown.
Regulations: Unknown.

Eureka, est. 01-02-1864, Yuma County, shown on Rand, McNally, 1866 GLO Map, USGS Report, ABM Map, and Hinton.
Boundaries: From Regulations: Beginning at Joe Johnsons Gulch on the Colorado River, then running up the river to Rode's Ranch and extending back at right-angles from the River at each of the above points 12 miles, thereby including all country within 12 miles of the Colorado River between Johnsons Gulch and Rodes Ranch.
Regulations published in King, 261.

Francis, est. ?, Coconino County, shown on USGS Report, ABM Map.
Boundaries: Unknown.
Regulations: Unknown.

Fort Rock, est. 01-27-1872, Yavapai County.
Boundaries: From Regulations: Commencing at a point twenty miles due east of the Fort Rock house and running due North for a distance of twenty miles, thence due west for a distance of forty miles, thence due south for a distance of forty miles, thence due east for a distance of forty miles, thence due north for a distance of twenty miles or to the place of beginning.
Regulations published in *Weekly Arizona Miner*, 3:3 (Feb. 17, 1872).

Globe, est. 11-25-1875, Gila County, shown on Rand, McNally, 1863 GLO Index, and ABM Map.
Boundaries: From regulations: Commencing at a point where the supposed line of the San Carlos Indian Reservation crosses Salt River; thence running down the course of said river to the mouth of Pinto Creek; thence in a southerly direction to the "Bloody Tanks" on the trail from the Globe mine to Pinal Post; thence along the summit of the timber range of the Pinal Mountains, to the Gila river; thence up the Gila River to the supposed line of the San Carlos Reservation; thence along said line to the place of beginning. This District is intended to embrace that portion of the San Carlos Reservation that is about to be cut off and declared open to the occupation of miners and prospectors. *See* Hinton, 142.
Regulations dated November 25, 1875, published in *Weekly Arizona Miner*, 1:3 (Dec. 17, 1875), King, 270; recorded in Book 1 (Misc.), page 173, Pinal County Records, regulations dated January 11, 1878, on GLO Microfilm.
Other names: Miami.

Gold Bug, est. ?, Mohave County, shown on USGS Report, ABM Map.
Boundaries: Unknown.
Regulations: Unknown.

Gold Mountain, est. ?, Pinal County, shown on USGS Report, ABM Map.
Boundaries: Unknown.
Regulations: Unknown.
Other names: Galiuro.

Golden Basin, 04-10-1880, Mohave County, shown on USGS Report, ABM Map.

Boundaries: From Regulations: Beginning at the Colorado River and the Eastern side of Sacramento Valley, thence Easterly along Southern Bend of Colorado River about 20 miles to the junction of said river and the Western side of Wallapai Valley; thence southerly along the Western side of Wallapai Valley to the intersection of the line forming due East and West through Mountain Springs, being a distance of about 35 miles, thence along said line about 20 miles to Eastern side of Sacramento Valley. Thence about 35 miles to point of beginning.

Regulations recorded in Book 1 (Misc.), pages 489–90, Mohave County Records.

Grand Canyon, est. ?, Coconino County, shown on USGS Report, ABM Map.

Boundaries: Unknown.

Regulations: Unknown.

Other names: Grandview.

Greaterville, est. pre-1864, Pima County, shown on 1863 GLO Index, USGS Report, and ABM Map.

Boundaries: From Regulations: Commencing at the point where the road from Thos. Gardiner's ranch joins the road coming down the Ophir Gulch from Greaterville; thence northerly to Mescal Springs; thence westerly to the northernmost point of the mine location known as the Catacomba, recorded in the records of the Smith District, folio 267; thence westerly to Alamos Ranch; thence south to T. Welisch's ranch; thence southeasterly to the site of Gardiner's "old" sawmill; thence easterly to the place of beginning, and shall include all the territory inclosed within those lines.

Regulations dated November 27, 1880, and February 19, 1881, on GLO Microfilm; and regulations dated January 5, 1893, on deposit with Arizona Historical Society Archives, Tucson.

Greenfield, est. 01-02-1905, Pima County.

Boundaries: From Regulations: Beginning at a place called the Fresnal Ranch, in the County of Pima, Territory of Arizona, in the eastern portion of the Comobabi Mountains, thence in a northerly direction a distance of about seven to ten miles to the 3rd Standard South Parallel; thence westerly along said Parallel a distance of about fifteen miles; thence southerly a distance of about twelve to fifteen miles to a point or place called Artesa; thence southeasterly a distance of about three to five miles to a place or point called Cholla; thence northeasterly a distance of about ten to twelve miles to the place of beginning.

Regulations recorded in Book 7 (Misc.), page 358, Pima County Records.

Greenlee Gold Mountain, est. pre-1864, Greenlee County, shown on 1863 GLO Index, USGS Report, and ABM Map.

Boundaries: From Regulations: Commencing at H. Lesinsky's dam about two miles above Clifton and running on a line Northwest to Eagle River, thence up Eagle River ten miles, thence to the mouth of the Bonita River, thence to Sunset Peak, thence to the place of beginning.

Regulations dated April 28, 1879, recorded in Book A (Mining Records), page 23, Graham County Records.

Other names: Metcalf.

Greenwood, est. pre-1878, Mohave County, shown on Rand, McNally, USGS Report, and ABM Map.

Boundaries: See Hinton.

Regulations unknown.

Other names: Signal.

Green Valley, est. ?, Gila County, shown on USGS Report, ABM Map.

Boundaries: Unknown.

Regulations: Unknown.

Other names: Payson.

Hackberry, est. pre-1876, Mohave County, shown on USGS Report, ABM Map.

Boundaries: Unknown.

Regulations: Unknown.

Other names: Peacock.

Harcuvar, est. pre-1866, La Paz County, shown on Rand, McNally, 1866 GLO Map, USGS Report, and ABM Map.

Boundaries: *See* 1866 GLO Map.

Regulations: Unknown.

Harquahala, est. ?, La Paz County, shown on USGS Report, ABM Map.

Boundaries: Unknown.

Regulations: Unknown.

Harshaw, est. 04-29-1878, Santa Cruz County, shown on USGS Report, ABM Map.

Boundaries: From Regulations: An area of twelve miles square taking the center of the French Mill Site as a center point.

Regulations recorded in Book 1 (Misc.), page 545, Pima County Records.

Hartford, est. 08-03-1879, Cochise County, shown on USGS Report, ABM Map, and Hinton.

Boundaries: From Regulations: This District is located in Ramsey Canyon in the Huachuca Mountains about 8 miles nearly south from Fort Huachuca, Pima County, Arizona. *See* Hinton.

Regulations recorded in Book 1 (Misc.), page 713, Pima County Records.

Hassayampa, est. 12-06-1863, Yavapai County, shown on Rand, McNally, 1866 GLO Map, USGS Report, ABM Map, and Hinton.

Boundaries: From Regulations: This district shall embrace all the ground from which the waters flow to the Hassayampa River, east of the eastern boundary of the Yapapi District, and north of the south-east corner of said Yapapi District, and shall be known as the Hassayampa District. *See* Hinton map.

Regulations published in *Weekly Arizona Miner* 1:2, 3 (Apr. 20, 1864), 1:2, 3 (May 25, 1865); and 4:1 (vol. 1?, 1867).

Helvetia, est. 04-16-1878, Pima County, shown on USGS Report, ABM Map, and Hinton.

Boundaries: From Regulations: This District shall embrace all the following lands having an area of ten miles square taking for a center point of starting the Santa Rita Chief Mine. *See* also Hinton map.

Regulations recorded in Book 1 (Misc.), page 530, Pima County Records.

Hope, est. pre-1881, Mohave County, shown on Rand, McNally, Hinton.

Boundaries: Shown on Rand, McNally, Hinton.

Regulations: Unknown.

Huachuca, est. pre-1881, Cochise County, shown on Rand, McNally, USGS Report, ABM Map.

Boundaries: Shown on Rand, McNally map.

Regulations: Unknown.

Hualapai (Wallapai), est. pre-1864, Mohave County, shown on Rand, McNally, 1863 GLO Index, USGS Report, ABM Map, and Hinton. Spelling "Hualapai" appears about 1891.

Boundaries: From Regulations: Commencing at Beals' Springs & running thence along the Prescott & Mohave toll road westerly ten miles, thence up the Sacramento Valley centeraly with the river range of mountains thirty miles, thence East twenty miles, thence down the Wallapi Valley to road & thence to Beal's Springs. *See* Hinton.

Regulations dated November 4, 1870, published in *Weekly Arizona Miner* 4:1, 2 (Dec. 2, 1871), and in King, 268.

Other names: Chloride, Cerbat, Mineral Park, Stockton Hill.

Humbug, est. pre-1881, Yavapai County, shown on Rand, McNally.

Boundaries: Shown on Rand, McNally map.

Regulations unknown.

Indian Secret, est. 06-28-1892, Mohave County, shown on USGS Report, ABM Map.

Boundaries: From Regulations: The Southern boundary to Commence at Louis Seabright house known as "Mountain Spring" thence in an Easterly direction to Dolan's Spring, thence in a northerly direction to the Centre of Salt Spring Wash to the bank of the Colorado River thence along the bank of the river to the mouth of the Sacramento Wash, thence in a Southerly direction along the wagon road to Louis Seabrights, the place of beginning.

Regulations published in Book 3 (Misc.), page 497, Mohave County Records.

Other names: White Hills.

Jacobs Lake, est. ?, Coconino County, shown on USGS Report, ABM Map.

Boundaries: Unknown.

Regulations: Unknown.

Johnson, est. ?, Cochise County, shown on USGS Report, ABM Map.

Boundaries: Unknown.

Regulations: Unknown.
Other names: Cochise.

Kofa, est. ?, Yuma County, shown on USGS Report, ABM Map.
Boundaries: Unknown.
Regulations: Unknown.
Other names: Humbug, Polaris.

La Paz, est. 10-06-1862, La Paz County, shown on 1866 GLO Map.

Boundaries are not described in mining district regulations, but the 1866 GLO Map indicates the district to be bounded by the Colorado River on the west, the Williams Fork District on the north, the Weaver Mining District on the south, and the Harcouver District on the east.

Regulations recorded in La Paz Records of Claims and Deeds, book 1, pages 168–69, Arizona State Archives.

La Fortuna, est. ?, Yuma County, shown on USGS Report.
Boundaries: Unknown.
Regulations: Unknown.

Laguna, est. ?, Yuma County, shown on USGS Report, ABM Map.
Boundaries: Unknown.
Regulations: Unknown.

Lone Star, est. ?, Graham County, shown on USGS Report, ABM Map.
Boundaries: Unknown.
Regulations: Unknown.

Little Cottonwood, est. 07-10-1882, Mohave County.

Boundaries: From Regulations: The centre of this District shall be the falls on Little Cottonwood Creek and extend from the falls five miles North, five miles South, five miles East and five miles West.

Regulations recorded Book 2 (Misc.), pages 718–19, Mohave County Records.

Other names: Renamed Richardson on November, 18, 1882.

Lost Basin, est. 04-06-1882, Mohave County, shown on USGS Report, ABM Map.

Boundaries: From Regulations: Commencing at the centre of Wallapai Wash where said Wash comes into the Colorado River and running from thence or nearly so Easterly along the South bank of the Colorado River to Pierce's Ferry on said River, thence up along the main wash and Wagon Road in a Southerly direction leading from Pierces Ferry to Grass Springs, thence Westerly to Pattersons Well (on the Wagon Road leading from Grass Springs to the Eldorado Mine), thence further west to the Eastern boundary of Gold Basin District and from thence Northerly along the Eastern boundary of Gold Basin District to centre of Wallapai Wash and place of beginning.

Regulations recorded in Book 2 (Misc.), page 701, Mohave County Records.

Lost Gulch, est. ?, Gila County, shown on USGS Report, ABM Map.
Boundaries: Unknown.
Regulations: Unknown.

Lynx Creek, est. pre-1881, Yavapai County, shown on Rand, McNally, ABM Map, and Hinton.
Boundaries: Shown on Rand, McNally and Hinton.
Regulations: Unknown.

Martinez, est. 03-08-1870, Yavapai County, shown on Rand, McNally, ABM Map.
Boundaries: See Rand, McNally map.
Regulations: Unknown.

Maynard, est. pre-1873, Mohave County, shown on Rand, McNally, USGS Report.
Boundaries: Shown on Rand, McNally map, Hinton.
Regulations: Unknown.

Meyers, est. pre-1881, Pima County, shown on Rand, McNally, ABM Map.
Boundaries: Shown on Rand, McNally map.
Regulations: Unknown.
Other names: Gunsight.

Miami. *See* Globe.

Mineral Creek, est. pre-1878, Pinal County, shown on Rand, McNally, ABM Map, and Hinton.
Boundaries: Shown on Rand, McNally and Hinton maps.
Regulations: Unknown.
Other names: Mineral Gulch, Ray, Kelvin.

Mineral Park, est. pre-1899, Mohave County, shown on USGS Report, ABM Map.
Boundaries: Unknown.
Regulations: Unknown.

Mineral Point, est. pre-1873, Yavapai County, shown on ABM Map.
Boundaries: Unknown.
No regulations found, although recorder's book was apparently filed with Yavapai County Recorder's Office during 1873.

Minnesota, est. 12-27-1880, Mohave County, shown on ABM Map.
Boundaries: From Regulations: Commencing at Johnsons Rock on the Colorado River and running east 20 miles, thence North 20 miles thence to Roaring Rapids, thence down the River to place of beginning.
Regulations recorded in Book 1 (Misc.), pages 514–15, Mohave County Records.

Mountain Spring, est. pre-1874, Mohave County.
No information other than location notice.

Montezuma, est. ?, Pima County, shown on USGS Report, ABM Map.
Boundaries: Unknown.
Regulations: Unknown.
Other names: Sonoita Mountains, Puerto Blanco Mountains.

Mowry, est. ?, Santa Cruz County, shown on USGS Report.
Boundaries: Unknown.
Regulations: Unknown.

Music Mountain, est. ?, Mohave County, shown on USGS Report, ABM Map.
Boundaries: Unknown.
Regulations: Unknown.

McConnico, est. ?, Mohave County, shown on USGS Report, ABM Map.
Boundaries: Unknown.
Regulations: Unknown.

Nogales, est. ?, Santa Cruz County, shown on USGS Report, ABM Map.
Boundaries: Unknown.
Regulations: Unknown.
Other names: Gold Hill.

Old Baldy, est. ?, Pima County, shown on USGS Report, ABM Map.
Boundaries: Unknown.
Regulations: Unknown.

Old Hat, est. 06-22-1878, Pima/Pinal Counties, shown on Rand, McNally, USGS Report, and ABM Map.

Boundaries: From Regulations: That the District be bounded as follows: to commence at Old Camp Grant on the San Pedro River that point to form the North East Corner of the District, thence following the Course of the River upst[ream] in a southerly direction Twenty-four miles thence in a westerly direction to the summit of the St Catheryn Mountain following the summit of the Mountain until the line strikes the Old Road from Tucson to Camp Grant thence following the Old Road to Camp Grant the place of beginning this mining District shall be known as the Old Hat Mining District.

[*Note*: It would have been important to designate a place of recording since the district overlapped two counties.]

Regulations recorded on July 29, 1878, in Book 1 (Misc.), pages 561–62, Pima County Records.

Other names: Oracle.

Oro Blanco, est. pre-1881, Santa Cruz County, shown on Rand, McNally, USGS Report, and ABM Map.
Boundaries: Shown on Rand, McNally and Hinton maps.
Regulations: Unknown.
Other names: Ruby.

Owen (Owens), est. 07-17-1874, Mohave County.
Shown on Rand, McNally, ABM Map, and Hinton.

Boundaries: From Regulations: Commencing at the southeast corner of Aubrey District and stream known as the "Big Sandy" and running down said stream to the mouth of the "Santa Marie" thence west fifteen miles thence northerly to the southwest corner of Aubrey District thence along the south line of Aubrey District to the place of beginning. *See* Hinton map.

Regulations recorded in Book 2 (Misc.), page 307, Mohave County Records.

Other names: McCracken, Potts Mountains.

Owl's Head (Owl Head), est. 01-18-1880, Pinal County, shown on ABM Map.

Boundaries: *See* Hinton.

Regulations recorded in Book 1 (Misc.), page 376, Pinal County Records.

Pajarita, est. 05-06-1880, Pima County, shown on ABM Map.

Boundaries: From Regulations: Commencing at the point on the Santa Cruz River where the river intersects the boundary line between Arizona and the State of Sonora, Mexico, thence following down the river to the old mill at the town of Tubac, thence to the peak known as the Oro Blanco Picacho, thence South 20 West to the boundary line between the Territory of Arizona and the State of Sonora, Mexico, thence along the said line to the place of beginning.

Regulations recorded in Book 2 (Misc.), pages 212–16, Pima County Records.

Palmetto, est. pre-1864, Santa Cruz County, shown on 1863 GLO Index, USGS Report, and ABM Map.

Boundaries: From Regulations: Commencing at a point on the Sonoita River at or near Sanfords House, thence down the Sonoita River to the junction of the Sonoita & Santa Cruz Rivers to the Sonora line, thence along the Sonora line to a pass between the Patagonia Mountains & the San AnnTone Mountains, thence along the summit of the Patagonia Mountains to a point on the Sonoita River about two Miles Easterly from the Old Astic Mill site, thence down the Sonoita River to place of beginning.

Regulations dated December 6, 1880, recorded in Book 2 (Misc.), page 441, GLO Microfilm.

Papago, est. 06-23-1873, Pima County, shown on ABM Map.

Boundaries: From Regulations: All that portion of the county of Pima situate about thirty-five miles in a southwesterly direction from the town of Tucson, in said county of Pima, bounded as follows: Commencing at a monument of stone five miles distant from the mouth of the main shaft of the Montezuma mine, thence north five miles, thence west ten miles, thence south ten miles thence east ten miles, thence north five miles to the place of beginning. *See* Hinton.

Regulations recorded Book 1 (Misc.), pages 234–36, Pima County Records, and published in *Arizona Citizen*, 4:2 (Jul. 5, 1873).

Other names: Sierrita.

Patagonia, est. pre-1881, Santa Cruz County, shown on Rand, McNally, USGS Report, ABM Map, and Hinton.

Boundaries: Shown on Rand, McNally and Hinton maps.

Regulations: Unknown.

Other names: Duquesne, Washington Camp.

Peck, est. 10-23-1875, Yavapai County, shown on USGS Report, ABM Map, and Hinton, 99.

Boundaries: From Regulations: Commencing at the mouth of Poland creek, at its junction with Turkey creek, thence running up Turkey creek to the mouth of Bear creek, to where Battle Flat creek empties into it; thence up the ridge, on the top of the same, between Bear creek and Battle Flat creek, to the top of Bradshaw mountain, at the head of said ridge, thence running due south to Poland creek, thence down Poland creek to the place of beginning.

Regulations published in *Weekly Arizona Miner* 2:2 (Nov. 12, 1875).

Pearce, est. ?, Cochise County, shown on USGS Report, ABM Map.

Boundaries: Unknown.

Regulations: Unknown.

Picacho, est. ?, Pinal County, shown on ABM Map, mentioned in State of Arizona Archives.

Boundaries: Unknown.

Regulations: Unknown.

Pilgrim, est. ?, Mohave County, shown on USGS Report, ABM Map.

Boundaries: Unknown.

Regulations: Unknown.

Pima, est. 03-24-1877, Pima County, shown on Rand, McNally, USGS Report, and ABM Map.

Boundaries: From Regulations: Commencing at a point two (2) miles due north from Castle Rock running thence east two (2) miles, thence south ten (10) miles thence west ten (10) miles thence north ten (10) miles thence east eight (8) miles to point of beginning. *See* Hinton.

Regulations recorded in Book 1 (Misc.), page 425. Rand, McNally shows Pima in position of Amole Mining District.

Other names: Olive, Twin Buttes, Mineral Hill.

Pinal, pre-1879, Pinal County, shown on Rand, McNally, referred to in location notice in 1879.

Boundaries: *See* Rand, McNally map.

Regulations: Unknown.

Pine Grove, est. pre-1864, Yavapai County, shown on Rand, McNally, 1863 GLO Index, ABM Map, Hinton, 99. *See also* Woolsey Quartz.

Boundaries: From Regulations: Shall include all that section of country known as the "Bradshaw Mountains," not included in any other lawfully formed mining district.

Regulations dated January 31, 1870, published *Weekly Arizona Miner* 1:1 (Feb. 25, 1871).

Other names: Woolsey Quartz, Crown King.

Pinto Creek, est. ?, Gila County, shown on USGS Report, ABM Map.

Boundaries: Unknown.

Regulations: Unknown.

Other names: Pinto Valley.

Piñon, est. ?, Pinal County.

Mentioned in State of Arizona Archives, no information.

Pioneer, est. 05-10-1863, Yavapai County.

Boundaries: From Regulations: The boundaries of this Quartz Mining & Mineral district are as follows: viz. Commencing at a Bald Mountain near the sink and to the Westward of Lynx Creek running in a Southerly direction following the dividing ridge of the waters of the Agua Frio and Hassayampa Rivers to a large Pine mountain about thirty-five miles in an Easterly direction from the place of commencement, thence in an Easterly direction to the Agua Frio River, thence up the Agua Frio River following the bed of the stream in a northerly direction to Woolsey's Ranch, thence to a westerly direction to the place of beginning. *See* Hinton.

Regulations in original district records on file in Yavapai County Records and published in King, 253.

Other names: Walker.

Pioneer, est. 07-11-1875, Pinal County, shown on Rand, McNally, ABM Map.

Boundaries: From Regulations: The northern boundary a line drawn east and west and ten miles long, the center of which shall be ten miles due north of Silver King mine; the southern boundary, a line drawn east and west ten miles south of Silver King Mine; the east and west lines shall be drawn north and south to connect the north and south lines, constituting an area of ten by twenty miles, with the Silver King mine in the center.

Regulations recorded in Book 1 (Misc.), page 9, Pinal County Records, and published in *The Arizona Citizen* 1:2 (Aug. 10, 1875).

Other names: Superior, Silver King.

Planet, est. ?, La Paz County, shown on USGS Report, ABM Map.

Boundaries: Unknown.

Regulations: Unknown.

Plomosa, est. 02-11-1879, La Paz County, shown on USGS Report, ABM Map.

Boundaries: From Regulations: The location of the boundaries lines of said District shall commence at Centennial Well, on the Ehrenberg and Prescott wagon road, and thence run in a westerly direction to Round Mountain about five miles west of Granite Wash; thence in a south easterly direction to the Eagle Tail Mountain; thence to the top of the Hacasilla Mountain; thence northwesterly to the place of beginning.

Regulations published in *Arizona Sentinel* 3:4 (Feb. 22, 1879); and 2:1 (Apr. 5, 1879).

Purtyman, est. 07-14-1913, Coconino County.

Boundaries: From Regulations: Center of the District to be at the Old Purtyman Ranch above the Falls on Oak Creek and extend fifteen miles in any and all directions.

Regulations recorded in Book 2 (Promiscuous), page 102, Coconino County Records.

Pyramid, pre-1865, Mohave County, shown on 1866 GLO Map.

Boundaries: As shown on GLO map.

Regulations: Unknown.

Quartz Mountain, est. 12-27-1863, Yavapai County.

Boundaries: From Regulations: Commencing at a Bald Mountain known as the North West corner of the Walker District running along the West line of said district to its South West corner. Thence in a North Westerly course along the Divide between the Waters of the Hassayampa and Granite creeks to the Granite Mountain. Thence in a straight line to the place of beginning.

Regulations recorded in original district recorder's book filed with Yavapai County Recorder's office, pages 11–15, published in *Weekly Arizona Miner* 1:2, 3 (Mar. 11, 1864), 5:1 (vol. 1 ?, 1867), and King, 260.

Quijota, est. pre-1879, Pima County, shown on USGS Report, ABM Map.

Boundaries: Unknown.

Regulations: Unknown.

Randolf, est. pre-1881, Pinal County, shown on Rand, McNally and Hinton maps.

Boundaries: Shown on Rand, McNally and Hinton maps.

Regulations: Unknown.

Red Rock, est. 04-30-1879, Santa Cruz County, shown on Rand, McNally as Red Cloud, USGS Report, ABM Map.

Boundaries: From Regulations: Commencing at the center Monument of the Kitty Mines which is situated on [indecipherable] about eight (8) Miles Southeast from the Souaita [?], Thence running east five miles; Thence running North ten miles, Thence running west ten Miles—thence running South ten Miles, Thence running East five Miles forming a district ten miles square.

Regulations recorded in Book 1 (Misc.), page 658, Pima County Records.

Other names: Red Cloud.

Reed, est. 04-18-1901, Pima County.

Boundaries: From Regulations: Commencing at a mound of stones five miles due North of said Caley's Peak, run thence East five miles to a monument of stones, being the N.E. corner of said District; run thence at right angles due South 10 miles to a monument of stones, being the S.E. corner of said District; run thence at right angles due West 10 miles to a monument of stones being the S.W. corner of said District; run thence due North at right angles 10 miles to a monument of stone, being the N.W. corner of said District; run thence at right angles due East 5 miles to the place of beginning.

Regulations recorded in Book 6 (Misc.), page 380, Pima County Records.

Richardson, est. 11-18-1882, Mohave County.

Reorganization of Little Cottonwood Mining District.

Boundaries: From Regulations: Beginning at a monument of stone and a stake erected on Mount Bennett, and marked "Mt. Bennett" distance about Six Miles East of Hackberry or nearly so, and running west to the Railroad Grade, thence South four miles; thence East ten (10) miles, thence north along the divide of the slate, range eight (8) miles; thence West ten (10) miles to the Railroad Grade, thence South to the point of beginning.

Regulations recorded in Book 1 (Misc.), pages 614–18, Mohave County.

Richmond Basin, est. ?, Pinal County, shown on ABM Map. Mentioned in State Archives.

Boundaries: Unknown.

Regulations unknown.

Rincon, est. ?, Pima County, shown on USGS Report, ABM Map.
Boundaries: Unknown.
Regulations: Unknown.

Rio Virgin, est. pre-1865, Mohave County.
No information, mentioned in location notice.

Riverside, est. ?, Pinal County, shown on USGS Report, ABM Map.
Boundaries: Unknown.
Regulations: Unknown.

Sacramento, est. pre-1866, Mohave County, shown on 1866 GLO Map.
Boundaries: Shown on GLO map.
Regulations: Unknown.

San Cayetano, est. ?, Santa Cruz County, shown on USGS Report, ABM Map.
Boundaries: Unknown.
Regulations: Unknown.

San Francisco, est. pre-1864, Mohave County, shown on Rand, McNally, 1866 GLO Map, USGS Report, ABM Map, and Hinton.

Boundaries: The boundaries of the district are not described in the regulations, but are shown on the 1866 GLO Map as being bounded on the west by the Colorado River, on the north by the Pyramid District, on the northeast by the Sacramento District with the southeast corner being at a point where the continuation of the eastern boundary reaches longitude 35 north, thence approximately south 75 west to a point of intersection with the Colorado River. Regulations on deposit with Mohave County Archives. *See also* Hinton map.

Other names: Union Pass, Vivian, Boundary Cone.

Santa Catarina (Santa Catalina), est. 11-23-1878, Pima County, shown on Rand, McNally, ABM Map.
Boundaries: Shown on Rand, McNally map.
Referred to in *Arizona Sentinel* 2:2–3 (Nov. 23, 1878).

Santa Maria, est. ?, La Paz County, shown on USGS Report, ABM Map.
Boundaries: Unknown.
Regulations: Unknown.
Other names: Big Horn.

Santa Rita Mountains, est. ?, Pima County.
Boundaries: Unknown.
Regulations: Unknown.

Santa Rosa Mountains, est. ?, Pima County, shown on USGS Report, ABM Map.
Boundaries: Unknown.
Regulations: Unknown.

Seattle, est. pre-1864, Yavapai County, shown on 1863 GLO Index.

Boundaries: From Regulations: Beginning at the mouth of the Rio Bonito, thence running west down the Rio Gila to Camp Thomas Military Reservation; thence running north to the Apache Indian reservation; thence running east along the line of the Indian Reservation to the Rio Bonito, thence south down the Rio Bonito to the place of beginning.

Regulations dated January 1, 1881, on GLO Microfilm.

Sedgwick, est. 11-20-1871, Yavapai County.

Boundaries: From Regulations: The following described lines is hereby made the boundary of the Dist now undergoing organization—From McClouds point east for a distance of eight miles & from thence south a distance of 12 miles thence west 12 miles, thence north 12 miles thence 4 miles east to McClouds point comprising a section of country 12 miles square.

Regulations published in *Weekly Arizona Miner* 3:2 (Jan. 1, 1872), and King, 270.

Seneca, est. ?, Yuma County.

Boundaries: Unknown.

Regulations: Unknown.

Silver, est. 01-20-1879, Yuma County, shown on Rand, McNally, USGS Report, ABM Map, and Hinton.

Boundaries: From Regulations: Commencing at the mouth of Charcoal Wash, on the Colorado River; thence into the Colorado River to Light House rock; thence in an Easterly direction to Ehrenberg Trail; thence Southerly to the Charcoal Wash, to the place of beginning.

Regulations recorded in Book 1 (Misc.), pages 303–4, Yuma County Records.

Silver Bell, est. ?, Pima County, shown on USGS Report, ABM Map.

Boundaries: Unknown.

Regulations: Unknown.

Silver Hill, est. 06-24-1890, Pima County, shown on USGS Report.

Boundaries: From Regulations: Commencing at the S.E. corner of Township 20 Range 10 S. Gila and Salt River Meridian—thence west six miles to SW corner of said township, Thence N. twelve (12) miles to adjoining Township corner, thence S twelve (12) miles to place of beginning.

Regulations recorded in Book 4 (Misc.), page 393, Pima County Records.

Silver King, est. ?, Pinal County, shown on USGS Report, ABM Map.

See Pioneer.

Smith, est. pre-1864, Pima County, shown on 1863 GLO Index, Hinton.

Boundaries: From Regulations: Commencing at Old Camp Cameron and running to a point four miles east of Gardner's Ranch; thence to Davidson Spring; thence to Sierrita; thence to Camp Cameron. *See* Hinton.

Regulations dated April 17, 1875, published in *The Tucson Daily Citizen*, 4:2 (Mar. 17, 1875).

Squaw Creek, est. ?, Yavapai County, shown on USGS Report, ABM Map.
Boundaries: Unknown.
Regulations: Unknown.

Stanley, est. ?, Graham/Greenlee County, shown on USGS Report, ABM Map.
Boundaries: Unknown.
Regulations: Unknown.

Summit, est. pre-1881, Pinal County, shown on Rand, McNally, ABM Map, and Hinton.
Boundaries: Shown on Rand, McNally and Hinton maps.
Regulations: Unknown.

Superstition, est. 12-30-1893, Pinal County, shown on ABM Map.
Boundaries: From Regulations: Boundary lines to commence at the Junction of the Verde and Salt Rivers and thence South to the base lines Gila and Salt River Meridian, thence East to the dividing Ridge between Marlins Ranch and the Fragien Canyon, Thence North to Salt River, Thence down Salt River to place of beginning.
Regulations recorded in Book 4 (Misc.), page 533, Pinal County Records.

Swansea, est. ?, La Paz County, shown on USGS Report, ABM Map.
Boundaries: Unknown.
Regulations: Unknown.

Swisshelm, est. pre-1881, Cochise County, shown on Rand, McNally, ABM Map, and Hinton.
Boundaries: Shown on Rand, McNally and Hinton maps.
Regulations: Unknown.
Other names: Elfrida.

Teviston (Tevis), est. ?, Cochise County, shown on USGS Report, ABM Map. BLM Map index cross-references Dos Cabezas under Tevis.
Boundaries: Unknown.
Regulations: Unknown.

Thumb Butte, est. ?, Yavapai County, shown on USGS Report, ABM Map.
Boundaries: Unknown.
Regulations: Unknown.

Tiger, est. 02-04-1871, Yavapai County, shown on Rand, McNally, USGS Report, ABM Map, and Hinton.
Boundaries: From Regulations: Includes all the country known as the "Humbug" from its head to a point ten miles down said stream. *See* Hinton, 99.
Regulations published in *Weekly Arizona Miner* 3:2 (May 13, 1871), and King, 269.
Other names: Harrington.

Tombstone (Tomb Stone), est. 04-05-1878, Cochise County, shown on Rand, McNally, USGS Report, ABM Map, and Hinton.

Boundaries: From Regulations: All of that portion of Pima County enclosed by the following lines, Commencing on the San Pedro River at a point opposite the old mines and about eighteen miles above the upper crossing of the San Pedro River. Thence East to the western base of the Dragoon Mountains. Thence southerly along the base of said Mountains continuing the line to the Mule Mountains. Thence West to the San Pedro River. Thence Northerly down said River to the place of beginning including all of that range of hills known as the Tomb Stone hills together with there [?]. *See also* Hinton.

Regulations recorded in Book 1 (Misc.), page 526, Pima County Records.

Truman, est. 02-11-1875, Pima County.

Boundaries: From Regulations: This mining district shall include the Santa Rita, Cayetano and Tubac Mountains and all outlaying spurs and foot hills of the same; the old, so-called Tumacacari Mission Grant and that portion of Territory lying between the Solero Ranch and the Santa Cruz River including southward to the Sonora River called and known as the Stuibaba.

Regulations recorded in Book 1 (Misc.), page 296, Pima County Records, and published in *The Arizona Citizen* 4:2, 3 (Mar. 13, 1876).

Tucson, est. 04-30-1878, Santa Cruz County.

Boundaries: From Regulations: Bounded as follows: On the east, by a line commencing at a point one mile and a quarter from the North Eastern base of the San Calletano Mountains thence running southerly along the Eastern base of the San Calletano Mountains to a point on the Sonoita Creek at or near Sanfords. Thence along the top of the Patagonia Range to San Antonia Pass on the south by the boundary line between Arizona and Sonora running westerly from San Antonio Pass to a point on the summit of the Pajero Mountains. On the west by a line running Easterly from said peak to the place of beginning.

Regulations recorded in Book 1 (Misc.), page 548, Pima County Records.

Turkey Creek, est. 07-10-1864, Yavapai County, shown on Rand, McNally, 1866 GLO Map, ABM Map, and Hinton.

Boundaries: From Regulations: Commencing at the high point of rocks at the head of Lynx, Bigbug, Turkey and Hassayampa Creeks, and from thence to and along the divide that separates the waters of Bigbug and turkey creeks. to the Agua Fria and down the Agua Fria to a point east of the Pine Mountain, in a southerly direction from the point of beginning and from thence west to the summit of the divide that separates the waters that flow into Turkey Creek from the southwest, and those of other creeks to the south and west, thence along said divide to the head of Cement Flat, and from thence along the northern edge of said flat to the divide that separates the waters of the main Hassayampa and that of the east fork of the same, and thence along said divide to the place of beginning.

Regulations recorded in original district records filed in Yavapai County, published in *Weekly Arizona Miner* 1:1, 2 (Aug. 24, 1864), King, 266.

Tyndall, est. 11-17-1876, Santa Cruz County, shown on Rand, McNally, USGS Report, ABM Map, and Hinton.

Boundaries: From Regulations: Commencing at the highest Santa Rita Peak that is to say the highest Peak of the Santa Rita Mountains in the County of Pima, Arizona Territory and run thence west twelve (12) miles, thence south twenty (20) miles, thence east twelve (12) miles thence north twenty (20) miles to the place of beginning to wit: the highest peak of the said Santa Rita Mountains. *See* Hinton.

Regulations recorded in Book 1 (Misc.), pages 394–97; meeting of November 26, 1877, Book 1 (Misc.), pages 516–17; meeting of December 28, 1877, Book 1 (Misc.), pages 607–08, Pima County Records.

Other names: Salero.

Verdi (Verde), est. 04-17-1876, Yavapai County.

Shown on USGS Report, ABM Map.

Boundaries: From Regulations: Commencing at a point on the Verdi River known as Rose Cañon running from thence West for ten miles thence south ten miles thence East ten miles thence North ten miles to place of beginning.

Regulations in original district records on file in Yavapai County Records.

Other names: Jerome Mining District after the discovery of the United Verde Extension 1910–20.

Vicksburg, est. ?, La Paz County, shown on USGS Report, ABM Map.

Boundaries: Unknown.

Regulations: Unknown.

Vulture, est. ?, Maricopa County, shown on USGS Report, ABM Map.

Boundaries: Unknown.

Regulations: Unknown.

Other names: Apparently carved out of Wickenburg District to include area of Vulture Mine.

Walnut Grove (Wagoner), est. 05-21-1864, Yavapai County.

Boundaries: From Regulations: . . . shall be bounded as follows, on the North by the Hassiampa District and on the East by the Agua Fria river running down the said river 30 miles or more thence West across to the Hassiampa to where the wagon road leaves that stream to Los Pimos and thence North from Los Pimos on the line of the Weaver Dist to the place of beginning–. *See* Hinton.

Walker Quartz, est. 11-24-1863, Yavapai County, shown on USGS Report, ABM Map. Formed from part of Pioneer.

Boundaries: *See* Pioneer.

Regulations in original district records on file in Yavapai County Records, and published in *Weekly Arizona Miner* 4:1 (Jul. 20, 1864) and 3:4 (Oct. 26, 1864), and King, 257–60.

Warm Spring, est. 04-16-1896, Coconino County, shown on ABM Map.

Boundaries: The regulations of the district do not contain an identification of the boundaries except that the township named the "Davis" township "be not less than six miles square."

Regulations recorded in Book 1 (Promiscuous), pages 80–81, Coconino County Records.

Warren, est. ?, Cochise County, shown on USGS Report, ABM Map.

Boundaries: Unknown.

Regulations: Unknown.

Other names: Bisbee.

Wauba Yuma (Waba Yuma), est. pre-11-11-1865, Mohave County, shown on 1866 GLO Map.

Boundaries: Shown on GLO Map.

Regulations deposited with Mohave County Historical Society. The meeting of miners held pursuant to notice dated November 11, 1865, suggest that the district had previously been formally organized. The records deposited with the Mohave County Historical Society are those that were filed with the Mohave County Recorder pursuant to the Territorial Mining Code and contained neither regulations nor a description of district boundaries.

Weaver, est. 03-20-1863, La Paz County, shown on Rand, McNally, 1866 GLO Map, USGS Report, ABM Map, and Hinton.

Boundaries: From Regulations: The District be bounded on the north commencing at a point on the River called the "Half Way House" situated between Olive City and La Paz, thence running Easterly to Las Posas thence running Southerly along the main arrotta to "Castle Dome" thence Westerly to the River thence running North by the River to the place of beginning.

Regulations published in King, 252.

Other names: La Paz.

Weaver, est. 02-12-1881, Mohave County, shown on USGS Report, ABM Map.

Boundaries: From Regulations: Commencing at the Southeast corner of Minnesota District at and to a spring South on said mountain down to a point known as Round Island on the Colorado River, thence up the River to a point known as Johnstons Rock, thence east to the starting point.

Regulations recorded in Book 1 (Misc.), pages 517–18, Mohave County Records.

Other names: Virginia.

Weaver, est. 06-25-1863, Yavapai County, shown on Rand, McNally, 1866 GLO Map, USGS Report, ABM Map, and Hinton.

Boundaries: From Regulations: Bounded as follows—to wit, commencing at the mouth or sink of the Hassayamp Creek following up the eastern bank of said creek to the Tanks on the Southern boundary line of Walkers, thence

West to the head of the Canyon of the St. Maria, thence southerly to Indian Springs continuing in said direction crossing Date creek near the Indian Cemetery ten miles from said crossing, thence east to the place of beginning.

Regulations published in King, 251.

Other names: Rich Hill.

Whetstone, est. ?, Cochise County, shown on USGS Report, ABM Map.

Boundaries: Unknown.

Regulations: Unknown.

Other names: Benson.

White Picacho, est. ?, Yavapai County, shown on USGS Report, ABM Map.

Boundaries: Unknown.

Regulations: Unknown.

Wickenburg, est. 05-21-1864, Maricopa County, shown on 1866 GLO Map, ABM Map.

Boundaries: From Regulations: This district shall embrace all the ground from which the waters flow to the Hassayampa River South of the Red Cañon situated ten miles South of Weaverville extended South to the White tanks, and shall be known as the Wickenburg District.

Regulations published in *Weekly Arizona Miner* 4:2 (Jul. 20, 1864), and King, 263.

Williams Fork, est. pre-1866, La Paz County, shown on Rand, McNally, 1866 GLO Map.

Boundaries: Shown on Rand, McNally map.

Regulations: Unknown.

Winchester, est. ?, Cochise County, shown on USGS Report, ABM Map.

Boundaries: Unknown.

Regulations: Unknown.

Woolsey Quartz, est. 09-30-1864, Yavapai County.

Boundaries: From Regulations: This district shall be known as the Woolsey Quartz Mining District, and shall be bounded as follows, to-wit: Commencing on the Agua Fria at the north-east corner of the Turkey Creek Quartz Mining District, thence following said boundary on the northern side to the high point of rocks at the head of Lynx, Big Bug, Turkey and Hassayampa Creeks, and from thence down the dividing ridge between the waters of Lynx and Big Bug creeks, and along said ridge until it makes down into the valley at or near the sink of Lynx creek, and from thence north-east to the main Agua Fria, thence down said stream to the place of beginning.

Regulations published in *Weekly Arizona Miner* 4:1, 2 (Oct. 26, 1864).

Wrightson, est. pre-1864, Santa Cruz County, shown on 1863 GLO Index, USGS Report, and ABM Map.

Boundaries: The original regulations of the Wrightson District are not available, but the regulations published in 1881 suggest that it included the Santa Rita Mountains.

Regulations dated March 13, 1881, published in *The Arizona Citizen* (Mar. 22, 1881), and in GLO Microfilm.

Yavapai, est. 09-28-1863, Yavapai County, shown on 1866 GLO Map, 1863 GLO Index.

Boundaries: From Regulations: That the District be founded on the North by commencing at the North end of Point of Mountain Range lying on the West side of the Assamp River, hear the headwaters of the Agua Frio River, from thence along the dividing ridge of said Mountain in a southerly direction to a point intersected by the Trail now traveled from People Ranch to what is known as the Tanks on the Assamp River, from thence along the said Trail in a Southerly direction to the South-East corner of what is known as "Webbers Ranch" from thence in a North West direction to Williams Fork River, from thence up the Main Branch of said River twenty (20) miles from thence to the place of beginning.

Regulations in original district records on file with Yavapai County Records, GLO Microfilm, and published in King, 257.

Yellow Stone, est. pre-1864, Cochise County, shown on 1863 GLO Index.

Boundaries: From Regulations: About eight miles north from Tres Alamos.

Regulations dated 02-21-1880 recorded in Book 2 (Misc.), page 68, Pima County Records, GLO Microfilm.

APPENDIX B

Full Text of Arizona Mining District Regulations

The following pages include the full text of Arizona mining district regulations as found in county records, published in newspapers, found in the records of the Bureau of Land Management or General Land Office, the 10th Census Records, or other sources. The source of these records, and an alphabetical listing of the districts, is found in appendix A. The following records are placed in chronological order by the date of the establishment of the district (or at least the date of the regulations or other estimate). Many of the recorded documents taken from county records are in longhand, and the writing is frequently unclear. *The text of the regulations has been copied as written (or deciphered with no attempt to correct misspellings or capitalization), but names are not so easily identified.*

1862.10.06

La Paz Mining District, Yuma County, October 6, 1862 (Published Book 1, La Paz Records, pp. 168–69. Arizona Archives Microfilm, 346.0448 L11)

At a meeting of miners and others interested in mining operations held at La Paz on the 6th day of October A.D. 1862, the following preamble to and regulations were unanimously adopted.

Preamble

Whereas it is essential to the security and protection of mining interests that they be held under laws governing the same, therefore it is hereby resolved:

1. That each quartz claim, whether it be of gold, silver or other metal, shall consist of 200 feet in length with all its dips, spurs and angles.

2. That each miner may hold by location one claim in each gold, silver or copper lode, the discoverer to hold an additional discoverer's claim in every lode he may discover.
3. Upon each claim, whether it be of an individual or a company, there shall be conspicuously placed a notice stating the day of the month on which it was located, the limits and direction of the same together with the name or names of claimants. Such notice to hold a claim for the term of 20 days from the date of location at the expiration of which period it neither worked nor recorded the claim shall be subject to forfeiture.
4. That a recorder, a resident of La Paz, be elected whose duty it shall be to record all claims presented to him for that purpose, such record to be made in a book kept explicitly for that purpose, that each notice presented for record shall be accompanied with a specimen of the ore from the claim shall recorded specimen to be numbered to correspond with the number of the claim on the record. Such specimen to be preserved for future reference a copy of the record concerning any claims to be furnished by the recorder free of charge on the application of any claimant thereof.
5. That the sum of one dollar be paid to the recorder for each claim recorded by each individual claimant.
6. The recorder to furnish the book, paper, etc.

1862.12.08

Castle Dome Mining District, Yuma County, December 8, 1862 (Shown on Rand, McNally, 1866 GLO Map, 1863 GLO Index, USGS Report, ABM Map. Regulations published in King, 247, regulations dated November 2, 1867, included in GLO Microfilm)

At a meeting held at La Paz on the 8th day of December 1862 by persons claiming interests in mineral veins near the Castle Dome range of mts Col Snively was requested to act as Chairman and H. Ehrenberg as Secretary of the meeting and the following resolutions were adopted.

That the District wherein said veins are situated be called the Castle Dome District and be bounded as follows: Beginning at the Peak known as Castle Dom—thence 10 miles south—Thence East 10 miles—Thence north 10 miles—Thence West 10 miles to the starting point.

2. That a mining claim in this District shall be 100 yards along said vein including all the angles spurs &c belonging thereto—
3. That the Discoverer or Discoverers of a vein shall be entitled to 100 yds extra on each & every vein discovered by him or them.
4. That in taking possession of claims the [boundaries] shall be clearly defined by conspicuous stake or monuments of rock with the names of persons claiming.
5. That 100 yards on each side of the vein where not conflicting with prior right shall be considered part of the claim, and shall belong to the same, with any and every substance or thing found within these bounds on or below the surface.
6. That all claims shall be recorded within 10 days after claiming them—
7. That all claims thus recorded shall be property described in their boundaries, and their relative position, as bearing and distances (where practicable) to any natural or artificial object stated.
8. That in consideration of the want of mining implements and materials in this section at present, the time for commencing operations on the claims is fixed to 15 March 1863.
9. That on and after that date all claims located shall be worked within 30 days.
10. That all claims shall be worked in good faith for at least 4 days in each month.
11. That companies holding various claims on the same vein shall not be obliged to work all of them severally, but that the working of any one of their claims in accordance with article 10 shall be sufficient evidence of good faith and ownership of the parties claiming.
12. No claims shall be considered abandoned or forfeited for suspension of work for want to water or on account of war Indians or any other unavoidable circumstance or obstacle.
13. A Recorder shall be elected for the District who shall record the different claims, as presented in a book kept for that purpose and he shall give certified copies thereof—
14. The fee for Recording shall be One Dollar for each claim, and no records of claims shall be made unless a specimen of the ore, coming from said claim shall accompany the statement to be recorded, which specimen shall be properly marked and preserved by the Recorder to serve as evidence in case of dispute hereafter.

15. The Recorder to be elected hereafter shall for the present keep his office in the town of La Paz Arizona—
16. Herman Ehrenberg is hereby chosen Recorder for Castle Dom District
17. Any five miners holding claims in this District shall have a right to call a general meeting of the miners interested in claims there, for the purpose of revising the mining laws, the election of Recorder or for any other purpose referring to the general interest of the District.
18. Any such meeting as mentioned in article 17 shall only be considered legal by having 2 notices thereof posted within the bounds of said district, and one at the Recorders office for at least 14 days previously.

H. Ehrenberg, Secretary
J. Snively, Chairman

Subsequent meetings

September 14, 1863: Labor requirements suspended until November 1, 1863.

November 1, 1863: Snively, Chair, and Geo. W. Leigh, Secretary. Changes summarized as:

Claims 300 feet along vein and 50 feet on each side

Discoverers entitled to extra claim of 300 feet.

Monuments 18 inches high.

Recording must be within 30 days. Recorder's fees are $2.50 per notice except if contains more than five name, $.25 per extra name.

Shaft or tunnel five feet deep required within 6 months to give possession for one year.

Recorder to be elected annually.

November 2, 1867: Oliver Lindsay, president, Thomas Bidwell, Secretary. Changes included:

Resolved 1st That the District shall be known as "Castle Dome" and bounded as follows—Commencing at the mouth of the Yuma Wash on the Colorado River, and thence in a direct line to the peak known as the Castle Dome & thence in a South easterly course to the Gila River, thence along said River to its confluence with the Colorado River thence along the main channel of the Colorado of point of beginning.

Resolved 2nd That Two Hundred feet shall constitute a claim, running with the Ledge or Lode and one hundred feet on each side with all dips spurs & angles of said Lodge or Lode—

Resolved 3rd Where parties owning mining claims have complied with section 25 of chapter 50 Howell Code & sunk a shaft ten feet deep on said mines are hereby declared valid and not subject to relocation. And all mining claims that have not complied with said section are hereby declared abandoned and subject to relocation. . . .

Resolved 7th . . . The Recorder shall not be compelled to make any Record without his fees in advance—

Resolved 8th The Recorder shall receive the following fees: Fifty cents for each name on a notice and twenty five cents for each certificate of location; fees in currency.

Resolved 9th That a shaft at least ten feet deep shall be sunk within twelve months from the date of the location.

Resolved 10th All Locations made subsequent to the passage of an act of Congress of the United States for the Government of Mineral Law and in conflict with said act is hereby declared abandoned and subject to relocation.

July 17, 1871 (Abraham R. Hoagland, Chairman, William P. Miller, Secretary):

Article I The jurisdiction of the following laws shall extend over all lode, lodge & vein mines and mining property within the following described and bounded District. Commencing at a point known and being the centre of the "Castle Dome Mine" on the "Buckeye" lode in the "Castle Dome" Mining Range, and running from said point North Ten (10) miles. South ten (10) miles, and all mines within said boundary shall be known as being in the "Castle Dome" Mining District.

Article II Any person shall be entitled to locate on any lodge or lode that is open to location by discovery or otherwise, one claim of Two Hundred feet with an additional claim of Two Hundred feet for discovery, & for each and every individual name located on same ledge forming one company Two Hundred feet; provided that there shall not be over three thousand feet held on any one ledge or lode by any one company and said claim or claims shall include all dips spurs angle & Variations of the ledge, and said claimants shall be entitled to one hundred feet of ground on each side of the lodge for the purpose of working said ledge.

Article III [30 days to commence work and sink a shaft 15 feet deep by three feet wide—this would hold for six months, 30 feet deep to hold for one year.]

Article V [Recorder's fees are $0.25 per claim of 200 feet]

Article VII Location shall be made by placing a good and sufficient monument at some point on the claim with a notice stating the number of claims and feet, names of parties date of location and the course of said claims from said monument.

January 9, 1875 (W. L. Hopkins, president, J. D. Rittenhouse, secretary):

2nd That the jurisdiction of all laws & regulations adopted by this meeting shall extend over all Lode, Ledge & vein mines and mining property within the following described and bounded district commencing at a point known as being the centre of the "Castle Dome Mine" on the "Buckeye Lode" in the Castle Dome Mining Dist. A.T. and running from said point North five (5) miles, and South (5) miles, East five (5) miles and West five (5) miles. The District to be ten (10) miles long and ten (10) miles wide, and all mines within said boundaries, shall be known as being the Castle Dome Mining District Yuma county A.T.

3rd [After January 9, 1875, the act of Congress passed May 10, 1872, is accepted as the law governing the mines of the District.]

5th That all records of locations of mines and mining property in the Castle Dome Mining District Yuma County A.T. from and after January ninth A.D. 1875 be made with the County Recorder of Yuma County A.T. within (30) thirty days from date of location.

1863.01.08

Colorado Mining District, Mohave County, January 8, 1863 (from the District Recorder's Book, pp. 1–7)

At a Meeting of the Miners of the Colorado District held at the San Juan Co. House in Eldorado Cañon January 8th 1863 to act upon the report of the Committee appointed by the Miners of said District at their Meeting of January 1st 1863. Dr C. W. Bush was called to the Chair and H. B. Coit chosen Secretary. The Report of the Committee was read and received, and the following Laws were adopted unanimously.

Sec. 1st This District shall be known as the Colorado District and shall be bounded as follows; Commencing at the Mouth of Eldorado Cañon running Ten miles North following the Course of the river, thence West Twenty miles thence South Twenty miles thence East Twenty miles to the River, thence North Ten miles following the Course of the River to the place of beginning.

Sec. 2nd The Officers of this District Shall consist of a President, Vice President and Recorder Elected by the Claim owners of this District who shall fill their Respective Offices for the term of one year from the date of the Election or until their successors are duly elected and qualified.

Sec. 3rd It shall be the duty of the President to call all Meetings of the Miners of this District when requested to do so by Five or more claim owners in writing of the District. To preside thereafter and discharge all duties of said Office.

Sec. 4th In the absence of the President from the District or his inability to act, it shall be the duty of the Vice President to act in his stead.

Sec. 5th If shall be the duty of the Recorder to keep in a suitable book or books a full and truthful Record of all publick meetings to place on Record all Claims brought to him for that purpose When such Claims shall not interfere with or affect the rights and interests of Prior locators Recording the same in the order of their date. It shall also be the duty of the Recorder to keep a duplicate Copy of the Book of Records open for the inspection of the Publick at all hours and to fulfill all other duties appertaining to said Office.

Sec. 6th The Recorder shall receive Two dollars and a half for Recording from One to Four claims and Fifty cents for each additional Claim when the distance is within Three miles of the Recorders Office. When over Three miles he shall receive in addition one dollar per Mile. He shall receive Two dollars for recording a Transfer or Deed not exceeding two hundred words. And for each additional one hundred words at the rate of Fifty cents per hundred. He shall receive Two dollars and Fifty cents for recording Tunnel Notices. He shall have the power to appoint a Deputy to act in his stead for whose Official acts he shall be held responsible and he shall hold his Office subject to the pleasure of the Recorder. It shall be the duty of the Recorder to place a Certificate of an appointment of a deputy Recorder on Record at the time of Appointment.

Sec. 7th All examination of the Book or Books of Record must be made in the presence of the Recorder or his Deputy.

Sec. 8th All Quartz Claims shall not be over two hundred feet in length. All previous located Claims shall be entitled to fifty feet of all Spurs or Offshoots extended from their junction with said Lode and Fifty each side of surface ground for the purpose of operating said Lode. The Discoverer or Discovered of a new Lode shall be entitled to one additional Claim for discovery. No Person shall hold more than one claim on the same Lode except as heretofore provided or by purchase.

Sec. 9th Notice of a claim or location of Mining ground by any Individual or by a Company on file in the Recorders office shall be deemed equivalent to a Record of the same. Provided it does not interfere with prior Locations. Such Notice shall be Recorded in the Book of recorder within Ten days of its being placed on file.

Sec. 10th Any Party shall be entitled to run a Tunnel or Cut from any favorable place through any Lode or Claim by giving the Mineral from said Lode to the parties owning the same. The owners or proprietor of any Tunnel Co. shall be entitled to all Lode they may find in said Tunnel that have not been discovered and located at the time of striking their Lode by having their Intersection duty Recorded.

Sec. 11th The Recorder shall go on the ground with any and all persons desiring to locate claims and shall see that proper Notices are placed on the ground and that Monuments or Mounds are erected showing the point of commencement of such Claims.

Sec. 12th A Claim shall hold for the term of Six Months from the date of Record and shall not be forfeited or subject to relocation during that time. After the expiration of said six Months, One day labor shall be put upon each and every Claim and said work shall hold the Claim for thirty days. Said labor shall be done within the Six Months aforesaid, Any party or parties owning Claims May do labor sufficient to hold a claim or claims a length of time at the rate of one days labor for each claim per month.

Sec. 13th Any Party or Parties owning a Claim or Claims shall be allowed to concentrate his or their labor upon any single Claim or set of Claims by running a Tunnel, Shaft or drift and if such labor is equal to the amount of labor required to hold his or their Claims on any Lode whatsoever it shall be deemed equivalent to work on said Claims, and they shall not be forfeited or liable to relocation. Provided That the Party or Parties shall before the expiration of each and every Month give to the Recorder a Certificate containing a list of all the Claims considered worked up. Said Certificate shall be Countersigned by one or more Person performing said labor, and shall state the number of days he had been at work on said Claim or Claims during the Month. Said Certificate shall be placed on file by the Recorder. Should any Party or Parties fail to give said Certificate his or their Claims where the actual labor has not been performed shall be liable to forfeiture and relocation. The Recorder shall receive twenty five cents for each Certificate so filed.

Sec. 14th Any Party or Parties expending funds for the purchase of Powder, Fuse, Drills and Mining tools that are in actual use on any of his Claims it shall be deemed as labor performed on his Claim or set of Claims. Provided a Certificate to that effect is filed with the Recorder as is required by Section 13th at the rate of Five dollars per day.

Sec. 15th Any Party or Parties owning a Claim or Claims not complying with the Laws of this District his or their Claims shall be forfeited and shall be liable to be relocated but the original date of the location and said Lode shall not be changed.

Sec. 16th Any Claim owner or owners doing the necessary labor to hold their Claim or Claims the labor so performed shall be considered as done on their Individual Claims and the balance not performing the necessary labor their claims shall be considered forfeited and liable to be relocated. Whenever work shall have been done upon a claim deemed to be of the value or cost of one hundred dollars or upward the claim on which such sums has been expended shall be deemed as belonging in fee to the owner of said claim and his assigns and the same shall not be subject to location or relocation by other parties afterwards, except by acknowledged abandonment by said owners of the ground, And when such labor is performed the Recorder of the District shall give said party a Certificate that such labor has been performed and said Certificate shall be countersigned by the President of the District.

Sec. 17th All Officers of this District may be removed from their Office by a vote of two thirds of the Resident Claim owners of the District, Provided that the Parties calling a Meeting of the Claim owners shall State in their call that the removal of an Officer or Officers of the District is one of the objects of the Meeting.

Sec. 18th All Claims located on or since the 9th day of October 1862 shall receive Ninety days grace in addition to the Three Months specified by the Laws under which they were located.

Sec. 19th No Person shall be allowed to vote or take any active part in a Meeting of the Miners of this District unless he shall be a resident Claim owner in said District.

Sec. 20th These Mining Laws may be altered or amended by a two third Vote of the resident Claim owners present of the District Voting for the same at any time when Ten days Notice of such intention shall be given.

Sec. 21st When a Meeting of the Claim owners is called, Notices shall be placed in Three of the most publick places of the District.

Sec. 22nd The Regular Election of the Officers of this District shall take place on the Second Thursday of each January. Said Election shall be by Ballot.

Sec. 23rd All Vacancies may be filled at any Meeting of the Miners by giving the requisite Notice.

Sec. 24th In the absence of any Officer qualified to administer Oaths the Officers shall qualify one another.

Moved and Seconded that the Laws go into force immediately and that all the Laws heretofore passed by the Miners of this District be repealed with the exception of the Law which appertains to the Recorder. Mr. Lewis elected June 1st 1862 he to fill his unexpired term. Unanimously adopted.

The Recorder signified his willingness to act upon and receive fees in accordance with the New Laws.

Motion to go into Election of President and Vice President. Adopted. Mr. Miller and Mr. Calez appointed tellers, for President, Dr. Bush received eight Votes and Mr. Wm. Calez Twenty Votes. Mr. Calez declared duly elected President for one year.

For Vice President Mr. Wilder received Nineteen Votes, Mr. Poindexter received Fifteen. Mr. Wilder declared duly Elected Vice President for one year.

Moved to adjourn, Carried

B. H. Coit Sec.

1863.05.10

Pioneer Mining District, Yavapai County, May 10, 1863 (Original book of recorder from Yavapai County Recorder's Office)

Miners Meeting on the Oolksipava River May 10, 1863

Mr. S. Shoup, President
J. V. Wheelhouse, Secretary

Names of original prospectors as required by Resolution No. seven—

Capt J. R. Walker
Jos. R. Walker Jr.
John Dixon
Jacob Linn
Jacob Miller
Jas. V. Wheelhouse
Jack Swelling [Swilling]
Frank Finney
S. C. Miller
George Blosser
A. C. Benedict
S. Shoup
T. J. Johnson
D. Ellis [Daniel Ellis Conner]
A. B. French
Chas Taylor
H. B. Cummings
Wm Williams
G. Gillahan [Gilliland]
Jackson McCrackin
Rodney McKinnon
Felis Cholet
M Lewis
Jas Chase
George Coulter

Preamble, laws, & resolutions adopted & passed by the "Walker" prospecting & mining company for their mutual guidance and protection at a meeting of said company on the Ookilipava River May 10th 1863

Sec 1st To all whom it may concern, be it known that the "Walker" prospecting & mining company have taken up certain portions Ookilipava river & tributaries for mining purposes have formed the said portion into a District to be called Pioneer District extending from the head of said river to a tree below the falls at the foot of the mountains (on which the notice of claimants is put up) taking in all tributarys, gulches, & ravines drained by said portion of river to main summit on both sides.

Sec 2d That at a miners meeting duly called & at which a majority shall pronounce pro & con, shall be in this Dist the manner by which all laws be made and adopted, disputes to rights of claims settled, extent thereof, litigation &c, & all other business appertaining to miners & their rights usually settled at said meetings in other mining Dists—

Sec 3 That there shall be one President to preside over said meeting & one Secretary, both of whom shall be elected by a majority of votes for the term of one year from the date of election—It shall be the duty of said Presd't to call all meeting & see that business belonging to said meetings be properly brought before it & freely discussed to the satisfaction of all parties concerned—The duty of the Secty shall be to keep a true & proper record (in writing) of the proceedings of each & all of said meetings—

Sec 4 There shall be a Recorder for said Dist appointed by the miners for the term of one year, whose duty it shall be to record all claims, sales & transfers of same & other transactions in said Dis't appertaining to said office in other mining district—He shall keep a true record of same in writing open at all proper hours for public inspection—

Resolutions padded & carried at the above meeting—

Resolved, That one hundred yards in length and fifty yards from each side of centre of gulch be considered a claim on this river—

Resolved, That each original prospector be entitled to one extra claim by right of discovery—

Resolved, That each member of the company having drawn by lottery the number of his claims shall have the privilege of exchanging one or both of them for any other unclaimed part of said river same dimensions as original being observed—

Resolved, That whereas but little time was taken to properly prospect, no claims be taken for persons outside of original prospectors until they, have definitely settled to which part of said river their claim may be exchanged to—

Resolved That the President be entitled to a fee of five dollars for each miners meeting which may be called to settle disputes or other individual business, to be paid by parties calling said meeting before the meeting be called & in case of winner suit to be refunded to Plaintiff & same around collected from Defendant—

Resolved That the names of all original prospectors be embodied in this document—

Resolved That Mr. T. J. Johnson be President of this Dis't

Resolved That Mr. Wheelhouse be Secretary

"That Mr. Wheelhouse be Recorder

"That the original claimants whose names appear in this document have nothing to pay for recording their first claims—

Resolved That the fee for recording claims to be fixed at Two dollars & fifty cents per claim—

Resolved, That no Mexican shall have the right to buy, take up or pre-empt a claim on this river, or in this Dis't for the term of six month, to date from the first date of June 1863 to Dec 1st 1863.

J. V. Wheelhouse Sect'y
S. Shoup President

Miners Meeting, Held on the Ookilipava River, June 10th 1863

Resolutions passed at said meeting—

Resolved, That the present boundaries of the Dis't be enlarged & its limits be extended to the Francisco River on the East, on the West to the divide of the river Aziamp & Antelope creek & include the Agua Frio river & its tributaries—

Resolved, That two days more "from date" be allowed Original prospectors for locating their claims—

Resolved, That each & all owners of claims mark & number them on trees or stakes at both ends so as to be properly understood—

Resolved, That claim holders shall have the right to say at which end of their claim or claims they shall commence measuring from when they are called upon to measure—

Resolved, That all claims taken up be recorded within ten days & no person allowed to take up claims for other when out prospecting—

Resolved, That all claims duly recorded be held for one year whether worked or not—

Resolved That 200 feet in lodes be considered a claim

Resolved, That Chinamen be excluded from working in any portion of this District—

Resolved, That all persons bringing in Mexicans or having them in their employ, record the names of said Mexican at the office of the Dis't Recorder—

Resolved, That persons employing Mexicans in any capacity be held responsible for all depredations upon property provided to have been committed by them—

Resolved, That upon the discharge of each such Mexican from employ notice be given the Recorder by the employer—

Resolved, That the fee for Recording each individual name be fixed at four bits—

Meeting of the 12th July 1863 Link Creek—

Resolved, That the law already passed with regard to Mexicans remain unchanged—

Resolved, That the "Walker party" or Original prospectors have the right to locate their discovery claims in any part of the Dist & have ten days from date to do the same—

Resolved, That no person hold more than one pre emption claim on each stream in this District

Resolved, That Asiatics & Sonoranians be excluded from working in this Dist—

Resolved, That a committee of three be appointed to measure & stake all claims on this stream, said committee to receive two bits per claim as remuneration—

Resolved, That the above named committee consist of Mr. Murray, Col Dobbins, & Jacob Miller.

Resolved, That a committee of three be appointed who shall decide who are & who are not Mexicans subject to the law of exclusion from taking up & holding claims in the Dist—

Resolved, That Mr. Green, Mr. Murray & Wheelhouse compose the above named committee—

J. V. Wheelhouse Secretary—
T. J. Johnson President

Resolution passed at above meeting not mentioned above—

That Recorders fees be reduced to $1.50 per claim for Recording and Companys claims as one & ten days from date in which to record the same be allowed to miners—

Meeting held January 10, 1864

Limited one placer claim by pre emption at one time (except for Walker party). Placer claims are transferred by bill of sale.

Meeting held March 6, 1864

The trial respecting a disputed mining claim between J.M. Sanford, Plf & McKinnie & Hulton Dfdts, same on to be tried before a jury of the whole.

The following witnesses were sworn & testified for Pltf, W. L. Griffin, H. L. Hubbard, J. Marsh, Crane, Ingalls, R. Kram & Lockhardt & the following witnesses were sworn & testified for Dfdt, S. Ruder, H. B. Cummings, T. J. Johnson & F. G. Gillihan. The case was argued & Jury gave as result

For Plaintiff Eleven (11)

"Defendants Eighteen (18)

& it was ordered that verdict for defendants be entered.

On motion the meeting was adjourned Sine Die

Van Smith Sec
By F. G. Christie

Meeting held March 14, 1864

[Addressed issues related to dangers "from the hostile Indians that surround us"] Resolution of May 10th 1863 in regard to Mexicans extended six months from date.

Meeting held on June 26, 1864

Resolved That all that portion of Walker Dist embraced on Turkey creek & its tributaries be struck off to be formed into another Dist at the pleasure of those holding claims in said section or locality

A.W. Adams Secretary
S. Shoop President

1863.03.20

Weaver Mining District, La Paz, March 20, 1863 (Recorded June 3, 1863 and original record printed in the *Morning Call*, San Francisco, April 9, 1863)

Notice is hereby given that a miners meeting will be held in Olive City on the 20th day of March for the purpose of considdering the importance of forming a new District to be composed of what is now known as the "Weaver Range" and other sections that it may be considered judicious to add thereto. Signed by request of many miners—

The meeting called in accordance with the above notice was organized by electing Mr. C. B. Green President and James S. Trimble Secretary.

The object of the meeting was explained by reading the preceding notice which had been duly posted throughout the district.

Mr. J. P. Newcomb moved that a committee be appointed to consider the division of the district and bring in resolutions in regard thereto, which motion was seconded and passed by the meeting. Some objections being raised it was reconsidered and the question left open for discussion. After some debate the sense of the meeting was taken directly upon the division of the district, which resulted in almost a unanimous vote in favor of division—

On motion a committee of five were appointed and elected by the meeting to draft laws for the district and also to define its limits, the following Gentlemen were elected to said committee—Jas Reed, J. B. Chevalier, W. B. Marshall, H. M. Oliver, Louis Robertson.

On motion the meeting adjourned for one hour to allow the committee time to prepare their report.

Meeting again convened at 2 oclock P.M. The committee upon the limits of the District, and resolutions for the government of the same, handed in the following report.

Olive City March 20th 1863

At a meeting of miners and others held at Olive City pursuance of notice it was

Resolved That the undersigned act as a Committee to draft resolutions and to fix the limits of a proposed new mining District, both to be submitted for the action of an adjourned meeting to be held at 2 o'clock of the same day—

The committee respectfully submit the following resolutions.

Resolved 1st That the district be known as the Weaver District

Resolved 2d That the District be bounded on the north commencing at a point on the River called the "Half Way House" situated between Olive City and La Paz, thence running Easterly to Las Posas thence running Southerly along the main arrotta to "Castle Dome" thence Westerly to the River thence running North by the River to the place of beginning.

Resolved 3d That any one may locate one claim only on each lode or ledge that may be discovered in the district and that in addition the discovered of any ledge shall have a discovery claim.

Resolved 4th That each claim both by location and discovery shall be two hundred feet on the ledge, and one hundred feet on each side along the ledge and following the dip of the ledge.

Resolved 5th That in locating any claim or taking one up by discovery, the party so locating or taking up shall place a conspicuous notice on said claim defining its boundaries & the direction in which it runs.

Resolved 6th Said notice shall hold the claim twenty days, after which time should there have been neither work done upon it, nor record made of it, such claim shall be subject to forfeiture.

Resolved 7th That each claimant do or cause to be done on his claim three day labor in every ninety days.

Resolved 8th That in case such claimant shall have done twelve days labor on his claim within six months from the time of commencement, such labor shall hold said claim for twelve months.

Resolved 9th That there be a Recorder elected for this District who shall be a resident of Olive City whose term of office shall be six months— It shall be his duty to record all claims presented to him for that purpose in a book kept for that special purpose, & that it be his duty should it be required to furnish a certificate of record to each one recording a claim, or transfer of claims or bill of sale—

Resolved 10th That for recoding such claim the Recorder be paid fifty cents and for recording deed or transfer two dollars and fifty cents— The records shall be open for inspection of any one who may wish to examine them in the presence of the Recorder or his deputy—

Resolved 11th That the above are not intended to have other than a prospective effect and consequently that they do not in any manner conflict or interfere with any interests held under the laws of the district of La Paz.

Resolved 12th That a certificate of record from the Lal Paz district for any claims now taken within the boundaries of Weaver district shall be recorded free of charge in Olive City—

The meeting then balloted for Recorder which resulted in the election of James S. Trimble. Fifty seven votes were cast.

The meeting then adjourned.

James S. Trimble Secretary
C. B. Green President

Meeting of April 2, 1863

Allowed the Recorder to reside in either Olive or Mineral City and extended the term of office to 12 months.

Meeting of January 11, 1865

On motion of Mr. Freman and seconded by Mr. Gird

Resolved That we adopt the Mining laws of this Territory subject to the action of Congress as the laws of this district hereby repealing all laws heretofore made and in force in this District

Resolved That we adopt the County Recorder as the recorder of this district

W Hanford Secretary
W. Thompson President

1863.06.25

Weaver Mining District, Yavapai County, June 25, 1863 (Published in 10th Census)

Name and Boundary of District.

1st This District shall be known as Weaver District and bounded as follows—to wit, commencing at the mouth or sink of the Hassayamp Creek following up the eastern bank of said creek to the Tanks on the Southern boundary line of Walkers, thence West to the head of the Canyon of the St. Maria, thence southerly to Indian Springs continuing in said direction crossing Date creek near the Indian Cemetery ten miles from said crossing, thence east to the place of beginning.

Size of Claims.

2d—The side of claims in this District shall be one hundred & fifty feet on creeks or Gulches and seventy five feet on each side.

Number of Claims &c.

3d No person shall hold but one claim in this District except the original discoverers (Ten in number) and the discoverer of new creek or Gulch diggings who are & shall be entitled to one additional claim all claims worked & Recorded within five days from the time of location shall hold good for sixty days. After the expiration of said sixty days all claims shall be worked on one day in ten.

Arbitration.

4th All disputes in reference to mining claims in this District to be settled by arbitration.

Mexicans.

5th No citizens of Mexico shall hold or work claims in this District except the boy Lorenzo Para who is one of the original discoverers, and should the miners employ any of the said citizens of Mexico, they will be held responsible for their good behavior, and should the said employer fail to comply with this article he or they shall forfeit all interest in the miners & leave the District.

Purchasing Claims.

6th No person or persons shall purchase or sell any claims in this District for sixty days from the adoption of these laws, nor shall any person take up and hold claims for non-residents of the District.

Recorder.

7th There shall be one Recorder elected whose duty shall be to record mining claims & bills of sale & preserve the laws of the district in a book or books to be kept for that purpose, said Records to be open for examination (free). His term of office shall be three months & until his successor is elected. He shall post or cause to be posted notice in three public places of the District notifying the miners of the expiration of his term, stating the time & place of election, and deliver to his successor all Books Papers and Maps belonging to his office. His fees for Recording shall be one dollar for each claim, and one dollar for each Bill of Sale. No Bills of Sale or claims will be valid unless recorded within forty-eight hours after date—

Calling Meetings.

8th Any five miners can call a meeting of the miners of the District by Posting notices in three public places in the District stating the object of the meeting, giving five days notice and signing their names to said notices.

June 25th 1863.
Arthur M. Henry Recorder

1863.09.28

Yavapai Mining District, September 28, 1863 (Book at Yavapai County Recorder's Office)

Leihys & Mellons Camp September 20th 1863
By Laws.

Notice is hereby given that a Miners Meeting will be held at Leihy & Mellon's camp on the 28th day of September for the purpose of forming a Mining District and to make laws governing the same—

September 28th 1863

The meeting called in accordance with above notice, was organized by electing A. B. Smith—President & Geo. W. Leihy Secretary—

A committee of 5 were elected to draft laws for government of District its boundaries—A. B. Smith, Geo. W Leihy, G. C. Welch, C. G. Mellon & G. C. Cross the gentlemen of committee

On motion the meeting adjourned one hour to allow committee to make report.

4 O'clock P.M. 28th 1863

Meeting again assembled—The committee made the following report & respectfully asked its acceptance & adoption as the laws & boundaries of the District.

Resolved 1st That the District be known as the Yavapai District.

2nd That the District be bounded on the North by commencing at the North end of Point of Mountain Range lying on the West side of the Assamp River, hear the headwaters of the Agua Frio River, from thence along the dividing ridge of said Mountain in a southerly direction to a point intersected by the Trail now traveled from People Ranch to what is known as the Tanks on the Assamp River, from thence along the said Trail in a Southerly direction to the South-East corner of what is known as "Webbers Ranch" from thence in a North West direction to Williams Fork River, from thence up the Main Branch of said River twenty (20) miles from thence to the place of beginning.

3rd That any one may locate one claim only on each Lode or Vein of Mineral that may be discovered in the District & that in addition the discoverer of any Lode or Vein shall have a discovery claim.

4th That each claim both by location & discovery be three Hundred feet on the Lode or Vein and One Hundred feet on each side following the dip of the Lode or Vein.

5th That on locating any claim or taking one up by discovery the party or parties so locating or taking up shall place a notice on said claim in a conspicuous place setting forth the direction in which such claim or claims are taken.

6th That said notice shall hold said claim thirty days from the time of putting up said notice, after which time if there is no record made of said notice the claim shall be deemed abandoned and subject to relocation by any other party.

7th That any person taking up or holding claims in this District neglects to work or does not cause to be worked for the time of Sixty days from the time of recording shall be deemed to have abandoned the same & the claim shall be subject to relocation.

8th that each claimant do or cause to be done three days work on each claim in every ninety from the time of taking up or locating their claim.

9th That in case such claimant shall have done or cause to be done twelve days labor within six months from the time of locating, it shall hold said claim for one year—

10th That there be a Recorder elected for this District who shall be a resident of the District, whose term of Office shall be six months or until his successor is elected, it shall be his duty to record all claims presented to him for record in a book kept by him for that purpose, & that it be his duty, should it be required, to furnish certificate of Record to each one recording a claim or transfer of claim or Bill of sale. That it shall be the duty of said Recorder to deliver all records of this District kept by him to his successor in office upon presentment of a duty certified certificate of his election by the Miners of the District.

11th That for recording each claim or notice of claim Fifty cents, and for recording deeds of transfer Two dollars and fifty cents. The Recorders book shall be open to the inspection of any person who may wish to examine in the presence of the Recorder or his Deputy. That the books, stationery &c. necessary for the office shall be furnished by the Recorder at his own expense—

12th The Recorder may appoint any suitable person resident of this District, to act as his Deputy—

The whole adopted.

The meeting then proceeded to ballot for Recorder; G. C. Welch declared unanimously elected for the next ensuring six months—

Geo. W Leihy Secretary
A. B. Smith President—
Adjourned sin die.

1863.11.24

Walker (Lynx Creek) Quartz Mining District, Yavapai County, November 24, 1863 (Published in *Arizona Miner*, July 20, 1864, 4:1, 2; *Arizona Miner*, October 26, 1864, 3:4)

[*Note*: The Walker Quartz and Placer Mining District was carved out of the Pioneer Mining District, and there was considerable overlap between the membership and rules. The first meeting was held on November 24, 1863, but it appears that many meetings may have involved the miners from both districts. At the November 24 meeting, T. J. Johnson tendered his resignation, and Captain Bogart was elected chairman; J. V. Wheelhouse was replaced by V. C. Smith as recorder. It was also obvious that some friction existed between the old Pioneer District and the new district in Lynx Creek. Most of this was probably a result of the fact that the foundations of the older Pioneer District kept changing the rules to their own benefit. A committee of V. C. Smith, Solomon Shoup, Cal Dobbins, Major McKinney, and Mr. Sandford were tasked to draft the quartz (lode) laws and Cal Dobbins, A. Thom, and Mr. McCrackin were to address placer mines. Their respective reports follow:]

To the officers and members of the miners meeting held Nov. 24th, 1863, at the office of Recorder Lynx Creek.

Gentlemen Your committee appointed to establish boundaries & draft Bye-Laws for the government of a Quartz Miners District would most respectfully submit the following Preamble Resolutions and Bye Laws for your consideration.

Preamble

Whereas owing to the increasing interest manifested by the Residents & Miners of this locality in Quartz & other Lodes containing metal of value—and to the many & various conflicting and untruthful reports that have originated and spread throughout the land concerning, and to the detriment of the Residents & Miners of this locality be it

Resolved, That we the residents & miners of Lynx Creek & vicinity will under any & all circumstances defend, protect, aid & assist any and all traders & persons whether citizens or not in the prosecution of right legitimate business while within this jurisdiction of our laws.

Resolved, That we denounce the originators of the many falsehoods, circulated by the faint hearted many who have returned to their shin warming firesides as a set unworthy the name of Pioneers & be it further Resolved, That the boundaries of this Quartz Mining & Mineral district be as follows viz. Commencing at a Bald Mountain near the sink and to the Westward of Lynx Creek running in a Southerly direction following the dividing ridge of the waters of the Agua Frio and Hassayampa Rivers to a large Pine mountain, about thirty-five miles in an Easterly direction from the place of commencement, thence in an Easterly direction to this Agua Frio River, thence up the Agua Frio River following the bed of the stream northerly direction to Woolsey's Ranch, thence in a Westerly direction to the place of beginning, and be it further

Resolved, That the name of this district be known as the Walker Quartz Mining District

Bye Laws

Art 1st The officers of this district shall consist of One president and one Recorder Whose term of office respectively shall be six months from the date of the election.

Art. 2nd It shall be the duty of the president to preside over and order all meetings, and to conduct the business of the same according to the rules adopted by legislative bodies

Art. 3rd It shall be the duty of the Recorder to act as secretary of all meetings and keep a true & correct record of all claims located and when required shall accompany the person or persons locating claims as a witness of the same & make out all deeds and transfers of claims and take acknowledgements thereto.

Art. 4th the Recorder shall receive for each name appearing upon the notice fifty cents and for each mile traveled by him when required in locating claims provided the same exceeds two miles travel from his office fifty cents per mile and for each deed or transfer the sum of $1.50. The Recorder shall procure the necessary books for his office & turn the same over to his successor.

Art. 5th The books of the Recorder shall be open at all times at his office for the inspection of the public.

Art. 6th The Recorder may appoint deputies who shall have the same authority, & whose acts shall be deemed as legal as the principal.

Art. 7th Notices of all meetings shall be posted one week previous to the meeting in three prominent places in the district which notices shall designate the time & place of holding the same.

Art. 8th All claims of Quartz or of lodes containing metal of value shall be two hundred feet along the lode with Fifty feet on each side & all the dips, spurs & angles following the ledge.

Art. 9th The discoverer of a ledge shall be entitled to one claim, which shall be known as the discovery claim, and one by right of pre-emption.

Art. 10th Person or persons locating claims shall post notices and erect monuments either by stakes ditches or furrows or stone or trees at each of his or their claims which notices shall designate the date of location, amount claimed, names of parties claiming the direction following the ledge as possible with its dips spurs & angles.

Art. 11th All claims shall be recorded within sixty days from the date of posting notices

Art. 12th Person or persons locating claims shall furnish the Recorder with a true copy of the notice and a specimen of the metal

Art. 13th No claim shall be deemed abandoned or subject to relocation after recording provided three days labor shall be performed on each claim every three months for one year, or if held by a company the name of days labor required for the number of claims held by them may be performed upon any individual claim held by such company upon that particular ledge on & after the 1st day of April next.

Art. 14th None but white persons shall be allowed to hold claims in this district.

Art. 15th All disputes with record to claims shall be settled by arbitration: The parties disputant each selecting an arbitrator the two arbitrators thus selected, selecting the third, the decision of the arbitrators shall be deemed final

Art. 16th No person or persons shall locate a claim or claims in this district who are non residents, or who are not within the district at the time of its location

Art. 17th No person or persons except the discoverer of a ledge shall hold more than one claim on each ledge discovered except by purchase.

Art. 18th All sales and transfers of claims must be made by deed properly acknowledged, and a note of the same entered upon the books of the Recorder, stating the date amount sold or transferred and to whom so sold or transferred.

Art. 19th Any person or persons or company holding claims who shall have complied strictly with the foregoing by laws for the term of one year from the first day of April next such claims shall be deemed and considered as Real estate & held as such

Art. 20th The amount of labor required to be performed upon each claim or claims for one year may be performed at one & the same time, provided the same is performed within three months from the first day of April 1864, or within three months after recording if recorded on and after said first day April.

Art. 21st Persons acting as arbitrators shall be entitled to a fee of $5 per day which fee must be deposited by the parties calling the arbitration with the President of the district.

Art. 22nd The person or persons losing the suit shall pay the expenses, and money deposited by the opposite party upon demand return to them by the President.

Art. 23rd All laws & part of laws conflicting with these by laws are hereby repealed.

Art. 24th The discovery claim upon each ledge is hereby except from the labor required in the foregoing by laws.

Meeting held on Lynx Creek at McLaughlin Store, September 16, 1865

Moved and Seconded that We as A boddy Are Empphaticaly Opposed to the Present Territorial Laws Carried Unanamously.

Moved and Seconded that A committee of 3 be Appointed by the Presadent to Draft A memorial to be Presented to the members of the Legislature form this District And Also presented to the Next meeting for its Adoption by the miners of Said District

Committee of 3 Appointed by the Presadent J. Brooks J. E. McCaffre J. Rees.

Meeting held on August 18, 1866

... J. E. McCaffrey moved and it was seconded that Recorder not be allowed to Record a claim until he goes on the ground and that he shall receive one Dollar per miles going and coming from the Lode in addition to the regular fee by law.

1863.12.06

Hassayampa Mining District, Yavapai County, December 6, 1863 (From *Arizona Miner*, April 20, 1864, 1:2, 3)

Mining Laws.

We propose to publish the laws of the several mining districts in this portion of the Territory. We begin with those of the Hassayamapa District of which Robert W. Groom, Esq., is Recorder.

Article 1. This district shall embrace all the ground from which the waters flow to the Hassayampa River, east of the eastern boundary of the Yapapai District, and north of the south-east corner of said Yapapai District, and shall be known as the Hassayampa District.

Art. 2. This District shall have a Recorder who shall hold office for one year from the time of his election, or until his successor shall be elected and qualified, whose duty it shall be to visit and examine either himself, or by his deputy, the ground claimed in any notice presented to him for record, before recording the same; and he shall be entitled to receive one dollar ($1.00) for each claim so examined and recorded.

Art. 3. A claim on any metallic vein or lode of quartz or other rock in this District, shall be three hundred (300) feet, running with the dips and angles of the lode, together with one hundred and fifty feet of ground on each side next to the lode, with all minerals therein contained.

Art. 4. All persons locating ground, for mining purposes, on any metallic vein or lode of quartz or other rock, in this District, shall be required to post a notice in a conspicuous place on the vein indicating as nearly as may be practicable its direction, and setting forth the number of feet claimed, each way from the notice; and the notice of the discoverer, or of the company, claiming the discovery claim as posted shall be the starting point from which any claims subsequently located on the same vein shall be measured.

Art. 5. No person shall be entitled to hold by location in this district, more than one claim on the same vein except the discoverer who shall be entitled to two.

Art. 6. Any notice claiming ground for mining purposes on any metallic vein or lode of quartz or other rock, in this district, posted according to the provisions of article 4th of this code, shall be deemed sufficient to hold such ground for the term of sixty days from the date of such notice; but if after the expiration of sixty days such notice be not found recorded, or filed with the Recorder for record, the ground so claimed shall be subject to re-location.

Art. 7. All notices claiming ground for mining purpose to any metallic vein or lode of quartz or other rock, in this district, properly located according to the provisions of this code shall be deemed sufficient to hold such ground until the first day of May, one hundred eight hundred and sixty-four; but if any claimant, or claimants, to any ground in this District, located and held as herein above provided, shall perform, or cause to be performed, the amount of six (6) days labor to each claim on any part of his or their ground, at any time between the date of the notice claiming such ground and the first day of May, 1864, the same shall be deemed sufficient to give him or them perpetual title thereto. The amount of labor, in all cases, to be estimated by the recorder, who shall on application, visit and examine the ground, and if in his opinion the required amount of labor has been performed thereon, he shall certify the fact in a note attached to or written beneath the notice of record in his office, claiming such ground, and for such service he shall be entitled to receive one dollar for each claim so examined.

Art. 8. Any person or persons holding ground for mining purposes, in this District, on and after the first day of May 1864, shall be required to perform, or cause to be performed, thereon the amount of three days labor to each claim in every thirty days, and if any person or persons shall comply with the above provisions for the period of one year, he or they shall thereby acquire perpetual title thereto, the labor to be examined and certified to by the Recorder, as provided in article 7th of this code.

Art. 9. If any person or persons holding ground for mining purposes in this District, who may perform or cause to be performed, thereon the amount of twelve days labor to each claim, at any time after the first day of May, 1864, he or they shall thereby acquire perpetual title to the same, the labor to be examined and certified to by the Recorder, as provided in article 7th.

Art. 10. A miners meeting may be called at any time in this District, by posting notices in three of the most public places, ten days previous to the term for which the meeting is called, signed by ten miners of the District, stating the object for which the meeting is called, and designating the place of holding the meeting.

Art. 11. All persons claiming mining ground in this District shall have a vote.

Art. 12. All laws or parts of laws heretofore in force in this District, in any wise pertaining to quartz mining are hereby repealed.

Art. 13. These laws shall be in force in this District from and after their passage.

Robert W. Groom, Recorder
Adopted Dec. 6, 1863.

Meeting on May 10, 1864 (Published in *Arizona Miner*, May 25, 1864, 1:3, 4)

At a regularly called meeting of the miners of Hassayampa District, Arizona Territory, held at Groom's cabin . . . called for the purpose of amending the laws so as to suit the convenience of all those who may desire to join the expedition soon to start east against the Apaches. . . .

Articles 7, 8 and 9 were amended to allow some flexibility in the performance of work.

Van C. Smith President
Robert McCoy secretary

1863.12.27

Quartz Mountain Mining District, Yavapai County, December 27, 1863 (Published in *Arizona Miner* on May 11, 1864, at 1:2, 3 and from Recorder's Book deposited with Yavapai County Recorder, 11–15)

Laws.

At a meeting of Quartz Miners held at Lount Cabin on Granite creek December 27th 1863. John West was chosen President and C. M. Dorman Secretary—

The following laws & regulations for the better governing of Quartz Mining, as reported on by the committee, were taken up separately and adopted.

Article 1st That this District shall be known & called the Quartz Mountain District, and shall be bounded and described as follows: Commencing at a Bald Mountain known as the North West corner of the Walker District running along the West line of said district to its South West corner. Thence in a North Westerly course along the Divide between the Waters of the Hassayampa and Granite creeks to the Granite Mountain, Thence in a straight line to the place of beginning.

Article 2nd That this District shall have a Recorder who shall hold his office one year from the time of his election, or until his successor be elected and qualified, and whose duty it shall be to visit examine & measure either himself or Deputy the ground claims in any notice presented him for Record before recording the same, & shall be entitled to receive one dollar for each claim so recorded and examined.

Article 3rd A claim on any Ledge or Lode of Mineral Rock in this District shall not exceed 300 feet, running with the main Lead, together with (150) feet of ground on each side of the man Lead or Lode with all the Minerals contained therein.

Article 4th All persons locating ground for Mining purposes on any Metallic Vein or Lode of Quartz or other Rock in this District shall be required to post a notice on a conspicuous place on the Vein indicating as nearly as possible the direction, & setting forth the number of feet claim each way from the notice, and the notice of the discoverer or the company containing the discovery claim shall be the starting point from which all claims subsequently located on the same vein shall be measured.

Article 5th No person shall be entitled to hold by location more than one claim on the same Vein, except the discoverer, who shall be entitled to two.

Article 6th Any notice claiming ground for Mining purposes on any Metallic Vein or Lode of Quartz or other Mineral Rock posted according to provisions of Art. 4th of this code, shall be deemed sufficient to hold such claim or claim for the term of sixty days from date of such notice. But after the expiration of sixty days, such notice be not found recorded or filed for record with the recorder of this district, the ground so claimed shall be deemed abandoned & subject to relocation.

Article 7th All notices claiming ground for Mining on any Metallic Vein or Lode of Quartz or Mineral Rock in this District properly located according to the provisions of Art. 4th of this code, shall be deemed sufficient to hold such ground until the first day of May One thousand eight hundred and sixty-four (1864) But if any claimant or claimants to any ground in this District, located & held as here in above described, shall perform or cause to performed the am't of (6) six days labor to each claim on any part of his or their ground at any time between the date of the notice claiming such ground and the first day of May 1864, the same shall be sufficient to give him or them perpetual title thereto Provided; that in case the claimant or claimants are not actually engaged in working the same at the expiration of two years from the time a perpetual title is acquired and certificate to that effect issued by the Recorder; the claimant or claimants shall renew the record in the Recorders Book provided for the purpose—Otherwise the claim to be deemed abandoned & subject to relocation—The amount of labor in all cases to be examined by the Recorder who shall on application visit & examine the ground, & if in his opinion the required amount of labor has been done there on, he shall certify the fact in a note to be attached or written beneath the notice recorded claiming such ground and for such services he shall be entitled to received one dollar for each claim so recorded.

Article 8th Any person or persons holding ground for mining purposes in this District, on and after the first day of May 1864, shall be required to perform or cause to be performed thereon the amount of three days labor to each claim in every ninety days, and if any person or persons shall comply with the above provisions for the period of one year, he or they shall thereby acquire perpetual title thereto, the labor to be examined and certified by the Recorder, as provided in art. 7th; Provided always that the labor be performed or record renewed according to Art. 7th of this code—

Article 9th If any person or persons holding ground for Mining purposes in this District perform or cause to be performed thereon the amount of twelve days labor at any time after the first day of May 1864, Shall thereby acquired perpetual title thereto; the labor be examined & certified to by the Recorder; And provided always that labor be performed or record renewed as provided in Article 7th of this code.

10th A miners meeting may be called at any time by posting notices in three conspicuous places in the District, ten days previous to the time of holding the meeting, stating in such notice the object for which the meeting is called, and place of holding the same, all such notices to be signed by the Quartz Miners of the District.

Article 11th All persons owning ground & residing in this District may vote at any meeting properly call by Quartz Miners. All persons non resident of this District may vote by proxy; provided they hold at the time of such meeting one claim in the District.

Article 12th The Recorder shall furnish at his own expense all books necessary for the recording of claims deeds of transfer & c. pertaining to his office, and keep the same open to the inspection of the public; and turn them over to his successor in office free of charge.

Article 13th An election shall be held on the first Monday in December of each year for the purpose of choosing a Recorder. A majority of the votes cast by ballot to elect. The Recorder so elected to commence his duties and take charge of the Books on the first Monday of January after this election.

Article 14th In case of the death or resignation of the Recorder, an election shall be ordered as provided in Art. 10th of this code for the choosing of a Recorder to fill his unexpired term of office.

Article 15th It shall be the duty of the Recorder to enter on the books for recording claims; underneath the notices, the name of the person who has examined such claim for record.

Article 16th All Laws or part of Laws heretofore in force in this District, pertaining in any way to Quartz Mining, are hereby repealed.

Article 17th These laws shall be in force from & after the date of their adoption.

On motion of Geo. Lunt—A. O. Noyes was put in nomination for recorder and unanimously elected.

We the undersigned committee appointed to draft and report By-laws & regulations for the better government of Quartz mining in Quartz Mountain District, having examined the foregoing Laws, do certify to them as a true & correct copy of the Original laws as adopted at the meeting called for that purpose Dec. 27th 1863.

(Signed) A. O. Noyes
E. M. Smith
Geo. Lount.

Meeting May 4th, 1864 (from page 27 of Recorder's Book)

[James Barney chosen as president and H. Brooks secretary.]

Whereas The Indians in this part of the territory are not at open war; And large numbers of them are lurking about the hills in this vicinity, watching for an opportunity to kill small parties of Whites—, making it unsafe for miners to work their claims, and furthermore, as many of the Quartz Miners of this district have been; and are intending to be for sometime to come, engaged in fighting he said Indians—it is therefore;

Resolved: That Article 7th, 8th & 9th of this code of Laws for the government of Quartz Mining in this District—, shall be and are hereby amended—So as to read the first day of October, in lieu of the first day of May. Thereby all right rights and privileges granted to Claimants by the said Code until the first day of May A.D. 1864 are extended to the first day of October 1864.

James G. Barney
H. Brooks, Secretary

1864.04.23

Cerro Colorado Mining District, Pima County
(Published in *Arizona Miner*, August 14, 1864, 1:2, 3)

At a meeting held this 23d day of April, A.D. 1864, at the office of the Director of the Cerro Colorado Mines, the following by-laws and regulations were adopted by the resident miners present:

Preamble.

The object of this meeting is to establish a Mining District, to be known as the District of Cerro Colorado, and to elect by ballot, in the usual form, a Recorder, whose duty it shall be to keep an office open from twelve to one o'clock of each day of week, (Sundays excepted) and to keep in said office a book of record of all mining claims presented to him for record, likewise all mortgages, transfers, deeds, and conveyances, and all other matters pertaining to said office of District Mining Recorder. The said recorder shall be entitled to receive for each claim presented to him for record the sum of one dollar and for each additional name in said claim the sum of fifty cents, to be paid upon presentation of said claim for record. The books of record and transfer to be open to the inspection of any person legally entitled to examine the same, upon the sum of one dollar to said recorder as fees. No books to be taken out of the office and must not be marred or defaced.

By-Laws.

Section 1. This district shall be known as the Cerro Colorado Mining District, to extend 12 miles east, 12 miles west, 12 miles north, and 12 miles south, from said Cerro Colorado mine, and shall be 24 miles square.

2. Each discoverer of a mine shall be entitled to a discovery share of 300 feet, in addition to his share of 300 feet, in each claim, not to exceed in any one claim 5,000 feet. The discovery claim to be exempt from assessment and sale except by the written consent of the said discovery claimant.

3. A discovery and location shall be entitled to 180 days from the date of record of such claim, to enable him or them to provide the necessary tools, provisions, etc., to commence operations, and thirty additional days of perform the legal requirements of the laws of the United states, and the laws and regulations of this district.

4. It shall be the duty of every discovery claimant to perform the following amount of labor within the prescribed time, and six days additional labor for each and every claimant, viz: to sink a shaft or tunnel thirty feet, exclusive of six days labor for each claimant. If the above conditions are performed, the claimant may hold said claim without any additional labor or expense, by recording a notice with said district recorder, for nine months from date of such record of intentions, and to suspend operation for said term of nine months.

5. After the expiration of the said nine months, if operations are not commenced, the said claim shall be considered abandoned and open for location anew.

6. Each claim shall be entitled to three hundred feet on each side of the vein or ledge, with all dips, spurs, angle or privileges, usually granted to mining claimants.

7. A meeting of the district can be called at any time after the expiration of thirty days from last meeting by the written notice of five members,

posted one at the recorder's office, one at Arevaca, and one at Soppho. Giving ten days notice of such meeting.

8. The above by-laws to remain in form until revised or altered by a two-third vote of a regular meeting, especially called for that purpose.

9. Each claimant shall post a notice of his or their discovery at the place of location, with the names of all the claimants written thereon at the time of location, with the date thereof, which notice shall remain in form for ten days, at the expiration of which time, it must be recorded or considered abandoned. No claim shall be considered perfected until recorded.

Wm Sim, Chairman
John A. Mahon, Secretary.

1864.05.21

Walnut Grove Mining District, Yavapai County, May 21, 1864, (Published in King, 263, shown on Rand, McNally, 1866 glo Map, and abm Map)

Laws & Regulations

Art 1. This Dist shall be known & called the Walnut Grove Quartz mining Dist & shall be bounded as follows, on the North by the Hassiampa District and on the East by the Agua Fria river running down the said river 30 miles or more thence West across to the Hassiampa to where the wagon road leaves that stream to Los Pimos and thence North from Los Pimos on the line of the Weaver Dist to the place of beginning—

Art 2d. This Dist shall have a recorder who shall hold the office for one year from the time of his election or until such time as his successor shall be elected & qualified & whose duty it shall be to visit, examine & measure either himself or his deputy the ground claimed in any notice presented to him for record, before recording the same & shall be entitled to receive one dollar for every claim so examined & recorded & shall also receive fifty cents (50c) per mile from the place of his residence to the place of the location to be recorded.

Art 3d A claim upon any ledge or lode of mineral rock shall not exceed three hundred feet (300) running with the main lode together with one hundred & fifty feet (150) of ground on each side of the main lode with all the mineral claim there in.

Art 4th All persons locating ground for mining purposes on any metallic vein or lode of quartz or other rock in this Dist shall be a sitizen of the United States & shall be to post a notice in a conspicuous place on the vein indicating as nearly as possible the direction & setting forth the number of feet claimed each way from the notice & the notice of the discoverer or the company containing the discovery claim shall be the starting point from which all claims subsequently located on the same vein shall be measured.

Art 5th No person shall be entitled to hold by location on the same vein except the discoverer who shall be entitled to two claims—

Art 6th Any notice claiming ground for mining purposes on any metallic vein or lode of quart or other mineral rock posted according to Article 4th of this code shall be deemed sufficient to hold such claim or claims for the period of sixty (60) days from the date of such notice, but if after the expiration of sixty (60) days such notice be found not recorded or filed for record with the Recorder of this Dist the ground so claimed shall be deemed abandoned & subject to relocation.

Art 7th All notices claiming ground for mining purposes on any metallic vein or lode of quartz or other mineral rock in this Dis't properly located in this Dis't according to the provisions of art 4 of this code shall be deemed sufficient to hold such ground & give them perpetual title thereto.

Art 8th A miners meeting may be called at any time for the purpose of amending or changing the laws of this Dis't by posting notices in three (3) conspicuous places in this Dis't ten (10) days previous to holding such election after publishing the same for three (3) months in any paper published in this territory stating in said notice the object for which said meeting is called & the place of holding the same all such notices to be signed by ten quartz miners of this Dis't—

Art 9th All persons owning ground & residing in this Dis't may vote at any election properly called. All persons non resident of this Dist may vote by proxy provided they hold at the time of such meeting one (1) claim in this Dis't.

Art 10th The Recorder shall provide at his own expense all books necessary for the recording of claims deeds, transfer &c pertaining to his office & keep the same open to the inspection of the public & turn them over to his successor in office free of charge—

Art 11—An election shall be held on the first Monday in December of each year for the purpose of electing a recorder a majority of votes cast by ballot to elect the same who shall commence his duties & take charge of the books on the first Monday in January after his election—

Art 12—In case of the death or resignation of the Recorder an election shall be ordered by posting three (3) notices in conspicuous places in this Dis't ten (10) days previous to the time of holding said meeting—

Art 13—It shall be the duty of the Recorder to enter on the books for recording claims underneath the notice the name of the person who examined such claims for record—

Art 14—All laws or parts of laws heretofore in force in this Dis't pertaining to quartz Mining are here by repealed—

Art 15—These laws shall be in force from & after the date of their adoption—

The above is a true copy of the laws in the Walnut Dis't as adopted & now in force—

Walnut Grove May 21st 1864—
F. M. Larkin Deputy Recorder

1864.05.21

Wickenburg Mining District, Yavapai County, May 21, 1864 (As published in *Arizona Miner*, July 20, 1864, 4:2)

Laws of the Wickenburg Mining District Arizona Territory.

Article 1st. This district shall embrace all the ground from which the waters flow to the Hassayampa River South of the Red Cañon situated ten miles South of Weaverville extended South to the White tanks, and shall be known as the Wickenburg District.

Art. 2nd This District shall have a Recorder who shall hold office for one year from the time of his election, or until his successor shall be elected & qualified & he shall be entitled to receive one dollar ($1.00) for each claim recorded.

Art. 3rd A claim on any Metalic Vein or Lode of Quartz or other rock in this District shall be three hundred feet (300) running with the dips, spurs & angles of the lode together with one hundred and fifty feet of ground on each side next to the lode with all mineral therein contained.

Art. 4th All persons locating ground for Mining purposes on any Metalic Vein or Lode of Quartz or other rock, in this district, shall be required to post a notice in a conspicuous place on the Vein indicating as nearly as may be practicable its direction, & setting forth the number of feet claimed each way from the notice, and the notice of the discoverer or of the company claiming the discovery claim, so posted, shall be the starting point from which all claims subsequently located on the same Vein shall be measured.

Art. 5th No person shall be entitled to hold by location in this District more than one claim on the same Vein, except the discoverer who shall be entitled to two.

Art. 6th Any notice claiming ground for mining purposes on any Metalic Vein or Lode of Quartz or other rock in this District, posted according to the provisions of Art. 4th of this code, shall be deemed sufficient to hold such ground for the term of thirty days from the date of such notice; but if after the expiration of thirty days such notice be not found recorded or filed with the Recorder for record, the ground so claimed shall be subject to relocation.

Art. 7th Any claimant or claimants to any ground in this district, located & held as herein above provide, shall perform or cause to be performed the amount of five days labor to each claim, on any part of his or their ground at any time between the date of the notice claiming such ground and ninety days thereafter shall be deemed sufficient to hold the same for the term of one year from date of record.

Art. 8th A Miners meeting may be called at any time in the District, by posting notice in three of the most public places (5) days prior to the time for which the meeting is called, by the recorder at the request of three miners of this district, stating the object for which the meeting is called and designating the time & place of holding the meeting.

Art. 9th No person shall be allowed to locate any claim on any Metalic vein or Lode of Quartz or other rock in this district for any non resident of this Territory.

Art. 10th All persons owning mining ground and recorded in this District shall have a vote—

Art. 11th Any Water privilege taken up in this District by notice & recorded, shall hold good for the term of one year from date of record.

Art. 12th All laws or parts of laws heretofore in force in this District, pertaining in any way to Quartz mining are hereby repealed.

Art. 13th These laws shall be in force in this district from & after there passage.

Adopted May 21st 1864.

Jas. A. Moore was elected recorder May 21st 1864

Jas. A. Moore Secretary
Henry Wickenburg President

1864.08.10

Turkey Creek Mining District, Yavapai County, August 10, 1864 (Published in *Arizona Miner*, August 24, 1864, 1:1, 2)

At a meeting of the claim holders of the Turkey Creek Quartz Mining Distrct, held August 10th, 1864, Charles Taylor was chosen President, and W. C. Collier Secretary. On motion it was voted that all of the old code of laws after the work "at" in the third clause of the first article be repealed, and the following laws be adopted in their stead, to be in force from and after their passage;

Article 1. This district shall be known as the Turkey Creek District, and shall be bounded as follows: Commencing at the high point of rocks at the head of Lynx, Bigbug, Turkey and Hassayampa Creeks, and from thence to and along the divide that separates the waters of Bigbug and turkey creeks, to the Agua Fria and down the Agua Fria to a point east of the Pine Mountain, in a southerly direction from the point of beginning and from thence west to the summit of the divide that separates the waters that flow into Turkey Creek from the southwest, and those of other creeks to the south and west, thence along said divide to the head of Cement Flat, and from thence along the northern edge of said flat to the divide that separates the water of the main Hassayampa and that of the east fork of the same, and thence along said divide to the place of beginning.

2. The officer of the district shall be a Recorder, to be elected by the claim holders of the district, and at stated elections, as hereinafter provided. He shall hold his office one year from the time of his election and qualification.

3. It shall be the duty of the Recorder to keep in a suitable book or books, a full record of al public meetings, to place on record all notices of claims brought to him for that purpose, recording the same in order of their date; to record all transfers, bills of sale, and deeds; to keep on file all original papers in his office until called for, and to call meetings for elections and other business of the district, by posting notices of the same as hereinafter provided. It shall be the duty of the recorder to keep his books of office open at all times for the inspection of the public; he shall also have the power to appoint a deputy to act in his stead, for whose official acts he shall be held responsible.

4. All examinations of the records must be made in presence of the Recorder or his deputy.

5. The fees of the Recorder shall be as follows: for recoding one claim

he shall receive the sum of one dollar, for transfer deeds, or bills of sale, one dollar, for each certificate of ownership he shall receive one dollars.

6. Each claim shall be three hundred feet on the lode, on a level, including all the dips, spurs, angles and variations of the same, with one hundred and fifty feet on each side of the lode, and following all its dips, angles and variations, and all the minerals contained with the same one hundred and fifty feet, (including all spurs or parts of spurs of the same lode) shall be considered as claimed by and belonging to the locator or locators of said main lode, and part of parcel of the same lode.

7. In locating a claim or claims notices must be posted in one or more conspicuous place or places, stating the name of claims so taken up and direction from discovery claim, or state the number of feet claimed and direction from a given point, at which point there must be a monument, erected, or a stake driven, or state what claim it joins, and also give the name of the claimant or claimants, and the date of the location of the same.

8. All claims shall be recorded within sixty days from the date of location and presentation of said claim notice to the Recorder, together with his fees for recording the same shall be equivalent to a record of the same. The time of the acceptance shall be endorsed on the back of the notice, and all claims not recorded in the time and manner hereinbefore specified shall be considered virtually abandoned and subject to re-location.

9. Any citizen of Arizona, minor heir, may hold any amount of claims by purchase, and one by location on each vein or lode, and the discoverer of a lode shall be entitled to one additional claim for discovery.

10. An election shall be held on the third Saturday in June of each year for the purpose of electing a Recorder. The election shall be held at a place within the district to be designated by public notice of the Recorder, posted ten days before the election, said notices to be posted in three public places within the district. In case the Recorder fails to post notices as above-stated in the proper time, any ten claim holders may call an election by posting notices in the same manner; and may also call an election to fill vacancies in the same manner.

11. Any person to be a legal voter, must be a claim holder in this district, but no vote shall be cast by proxy.

12. A certificate of ownership must be demanded within six months from the date of the record of the notice, claimant the ground for which said certificate is demanded. But no person shall be entitled to a certificate until he or she shall have been on the Lode and claimed his or her

location, in person. No one shall have a right to demand a certificate until the expiration of twenty days after the notice shall have been recorded for which said certificate is demanded. But if a certificate is not demanded in the time and manner as heretofore specified, his or her right to demand a certificate shall be considered as forfeited and the ground so claimed subject to re-location by any other person; provided, that sickness, inability, unavoidable accidents or detention in the service of armies of the United States has made it impossible for them to comply with the provisions of this law, they shall retain the right to demand and obtain a certificate within a reasonable time after their inability has been removed. Any certificate illegally or fraudulently obtained may be invalidated by a miner's meeting called for the purpose, and a certificate granted to the legal claimant. The Recorder is enjoined from granting any person a certificate until they will have complied with the letter and spirit of the laws of the district. When a certificate shall have been legally acquired the title so granted shall be irrevocable by any subsequent action of miner' meetings.

13. All disputes relative to the ownership of ground, claimed under the laws of this district must be settled before a certificate can be demanded. Either party to a dispute may call a miner's meeting, and by giving the other parties notice of the fact ten days previous to the meeting, the decision of said meeting shall be final.

14. All called meetings of the voters of the district must be done by posting notices (signed by the claim holders of the district) in three conspicuous places in the district, stating the object for which said meeting is called, and said meeting shall not have any power to transact any other business except such as was named in the posted notice, calling said meeting.

15. Discoverers shall have the right to locate the discovery claim on any part of the lode, provided that they locate it before any other claims have been located. All claims must be number from the discovery claim.

16. these laws may be amended or repealed by a two-third vote of all voters present at a meeting of the miners that shall be been called for that purpose, provided that twenty voters be in attendance and vote.

17. These laws shall not interfere with any rights acquired under the old laws

18. The recorder shall act as Secretary at all miners' meetings, if present. The voters may elect a President pro tem.

Chas. Taylor, Recorder

Meeting of January 28, 1865 (Recorded at pp. 9–11)

[T. J. Arnold chair and Henry Clifton secretary]

1st Resolved, that the Territorial mining law shall govern the mines of this District, provided that each claim or pertenencia shall be numbered each way from the discovery claim 1, 2, 3, &c. North South East or West (as the case may be) each claimant holding the ground on the vein corresponding the number opposite his name.

1864.09.14

Bradshaw Mining District, Yavapai County, September 14, 1864 (Published in the *Arizona Miner*, October 26, 1864, 1:1, 2, shown on 1866 GLO Report, and King, 264)

Montezuma City, Sept. 14, 1864

At a meeting of the miners of the Bradshaw Mining District, on Turkey Creek, in the Territory of Arizona, Mr. Moore was called to the chair, and Dr. G. M. Willing appointed Secretary. The following proceedings were then had:

The district was declared to be ten miles square, commencing at a monument in the town of Montezuma, on Turkey Creek, which shall be the centre of the district; and this in open meeting was declared the limits of the boundaries of the said mining district of Bradshaw.

On motion Mr. M. Solomon was nominated and by acclamation elected President of said mining district for the time specified in its laws. Dr. George M. Willing was nominated and duly elected Recorder.

Laws.

Section 1. The name, style and title of this district shall be the Bradshaw Mining District of Arizona Territory.

2. The officers of his district shall be a President and a Recorder.

3. It shall be the duty of the President to preside at all meeting of the miners, and cause the laws, rules and regulations of the district to be faithfully executed, and until the government of the United States shall, by act of Congress or otherwise, establish laws regulating the mines. The President shall be ex-officio judge of the miner's court, before whom all cases relating to the mines shall be tried. He shall hold his office for the term of one year, or until his successor is duly elected and qualified. In the absence or inability of the President the Recorder shall perform the duties pertaining to the office of President.

4. It shall be the duty of the Recorder to keep a true and correct record of all mining lots, claims, discovery lots, and building sites, or lots or parcels of ground not enumerated herein, in a book set apart for that purpose. He shall also in person or by deputy bound and number all mining claims or lots or parcels of ground, as set forth in these laws, and give to the claimant a certified copy of the same on payment of fees. He shall also be ex-officio clerk of the miner's court.

5. The said Recorder, for compensation for such services, shall be entitled to the following fees: For recording a placer, ledge or other claim $1; for recording mill site or lot, $5; recording deed or transcript, $2.50; affixing seal of office, 25 cts.; administering an oath, 25 cts.; each summons and subpoena, 2 cts.

6. On the 14th day of September, 1865, and on the 14th day of September of each succeeding year, there shall be held an election for the officers of this district, which election shall be by ballot—Any five persons present & qualified voters under these laws rules & regulations may appoint a Judge of said election which Judge shall appoint two clerks & proceed to receive the notes of the miners for the several offices to be elected. No person shall be a voter at any such election after the 14th day of September 1864 unless such person shall have resided in this District thirty days prior to the day of said election—And no person shall be entitled unless such person is connected or associated with same mine in this Dist & and provided further that no person shall be entitled to vote by proxy—

7. It shall be the duty of the said clerks of each election held within the district to record the names of each voter and mark upon the ticket the number of such ballots in plain figures, each clerk keeping a separate list of names and votes and at the close of the ballot box to record the notes counted by the judges of said election in the presence of three witnesses, one copy of which record, of the names and votes, shall be filed with the Recorder, and one copy with the President of said district.

8. Each person within this district shall be entitled to hold the following claims: The working of one of said claims shall be evidence of the working of all, and shall hold the same as if all were worked, provided, however, that each and every discovery lot or claim and one pre-emption lots or claim, by location, near said discovery, shall be held inviolate to the discoverer, whether worked or not, or whether it shall be one or the other of the claims mentioned in this section: One discovery lot on all ledges, and one by location, one placer claim, one hill claim, each consisting of three

hundred feet square, and one lot for building purposes, 50 by 150 feet, and one water claim 300 yards square, and one mill site one hundred yards square.

9. No person shall take up a claim within this district for agricultural or ranching purposes to the exclusion of mining operations, nor shall such claimant to such land hold the same only under the laws of this mining district, and the same may be taken up as mining claims by the miner thereof. Provided, however, that all improvements of value on such land shall not be interrupted or damaged, but when a valuable mine is known to exist and any such improvements be thereon, the damages shall be assessed by three disinterested persons, and the claimant of said mine shall pay to the owner of such improvements the assessed value of such damages.

10. Two or more persons may form themselves into a company for the more profitable and perfect working of their claims, and said company or companies shall be protected and defended as is provided for each and every individual miner.

11. Said company or association may taken up and hold a claim of one of all the above mentioned classes of claims for each and every member of said company, and for each operative in their employ, and that may be duly employed by said company in the working of said claims. Each company shall have the right of all the discoveries made by them as single miners, with right of way for roads, sluces, mills, tunnels or other operations necessary for mining purposes. Any miner or company of miners may, in addition to the grants specified in these laws, acquire and retain by transfer, devise, or purchase, any claim or claims by any miner of this district or persons holding a claim therein, and said transfer shall be by deed conveyed and acknowledged before the record or said district, and recorded by him. And when the party resides without the district, then such transfer shall be acknowledged before any judge of any court of record in the United States or the Territories, and the same shall be recognized by this district as valid and admitted to record.

12. These laws may be altered or repealed only in the following manner: Any eight persons, members of this district, may petition to the President thereof to call a special meeting of the miners, or at their regular sittings, stating for what purpose in full, and upon the receipt of such petition the President shall cause six notices to be put up in six of the most public places of the district, ninety days before the day of calling together the miners for such meeting, to take into consideration only the subject matter contained in said petition; and the same rules shall be applied to all such meetings as

in applicable to voters in this district, each member thereof to be a qualified voter; and then it shall be determined whether the matter contained in the such petition shall be acted on. And no other matter or subject relating to the miners shall be acted upon in said meeting, unless two-thirds of the miners of said district shall consent to the same, and are present at said meeting & provided further that such alteration change or repeal of any law rule or regulation shall not affect in any manner whatever any claim or claims or right, of any description whatever owned or claimed by any person or persons under the pre-exiting laws of this Dist prior to the said alteration, change or repeal of the same—

13. Every miner shall be considered a representative of this district, and shall be entitled to a seat in said meeting.

14. It shall be the duty of the clerk to keep a correct list of names of such miners alphabetically arranged, and call the names as if it were a regularly constituted legislative body, and when it is required by any member to note absentees. The Recorder shall power to appoint a clerk when it is his duty to preside.

15. Parliamentary usages shall be observed in all miners meetings, and such rules as the miners shall adopt.

16. It is further provided that on account of the danger apprehended by depredations from Indians, and the safety of the mining community, no person or persons shall be compelled to work any mine for the space of one year, commencing on the 14th day of September, 1864, and ending September 14, 1865.

17. These laws, rules and regulations shall be in fore from and after their passage.

Max Solomon, President
s/ G.M. Willing, Sec'y

1864.10.02

Woolsey Mining District, Yavapai County, October 2, 1864 (As published in *Arizona Miner*, October 26, 1864, 4:1, 2)

At a meeting of the quartz miners of Big Bug community held on Friday, September 30, John Roberts, Esq., was called to the chair and James O. Robertson elected Secretary. A new mining district was formed and the following by-laws were adopted:

Section 1. This district shall be known as the Woolsey Quartz Mining

District, and shall be bounded as follows, to-wit: Commencing on the Agua Fria at the north-east corner of the Turkey Creek Quartz Mining District, thence following said boundary on the northern side to the high point of rocks at the head of Lynx, Big Bug, Turkey and Hassayampa Creeks, and from thence down the dividing ridge between the waters of Lynx and Big Bug creeks, and along said ridge until it makes down into the valley at or near the sink of Lynx creek, and from thence north-east to the main Agua Fria, thence down said stream to the place of beginning.

2. This district shall have a Recorder, who shall be elected by the quartz mines of this district, and whose term of office shall be for one year from his election and qualification.

3. It shall be the duty of the Recorder to keep in a suitable book a full and true record of all notices or claims brought to him for that purpose recording the same in order of their date. To record all transfers, bills of sale, mortgages, conveyances, and deeds, and to keep on file all original papers in his office until called for by the proper owners.

4. It shall be the duty of the Recorder to keep the books of his office open at all times for the inspection of the public. He shall have the power to appoint a deputy, who shall have full power to act in his stead, and also for whose official acts the Recorder shall be responsible.

5. All examinations of the records must be made in the presence of the Recorder or his deputy.

6. The Recorder shall provide at his own expense all books necessary for the recording of claims, transfers or deeds, and for issuing certificates of title, and turn the same over to his successor in office free of charge.

7. The Recorder shall be entitled to certain fees, to-wit: For each claim recorded he shall receive one dollar. For transfer of deeds, or bills of sale, one dollar and fifty cents. For each certificate of title, fifty cents.

8. Each claim shall consist of two hundred feet on a level, including all dips, spurs, angles and variations, with 50 feet of ground on each side of the ledge or lode, and following all the dips, spurs and angles, with all the mineral contained therein.

9. All persons locating ledges of quartz or other rock for mineral purposes in this district shall be required to post a notice in a conspicuous place at the beginning of the ground claimed setting forth the direction claimed as nearly as possible, also the number of claims and the names of the claimants, and the total amount claimed. The locators are required to post notices and erect monument at the two extreme ends of the amount clamed.

10. All persons locating claims will be required to certify to the Recorder that each claim has been duly located according to the laws of this district.

11. All claims must be recorded within thirty days after location, or such claims shall be declared abandoned.

12. No person will be entitled to hold by location more than one claim on the same ledge or lode, except the discoverer, who shall be entitled to one additional claim for the discovery.

13. For all claims located after the first day of October, 1864, each company locating ground together will be required to sink a shaft to the depth of ten feet, and of sufficient width to describe the size of the ledge. Said shaft must be slunk within sixty days after location of the claim.

14. After the labor prescribed in the foregoing section shall have been performed the claimants of the ledge on which such labor has been done may demand of the Recorder (after his knowledge of that fact) a certificate of title, which certificate shall give a title for six months from date of labor performed. If after the expiration of six months operations are not commenced within thirty days said claim shall be declared abandoned.

15. All claims must be monumented from the discovery claim, in whichever direction they may lay from the discovery.

16. All claims that may have been located within the boundaries of this district before the formation of this district shall be subject to these laws, and all labor that may have been performed previous to that time, shall be credited to the parties who performed such labor.

17. Any citizen of Arizona shall be entitled to hold mining ground within this district by location, subject to the laws of the district.

18. An election shall be held on the first Saturday in October in each year for the purpose of electing a Recorder. The election shall be held within this district, to be designated by notice posted by the Recorder ten days previous to the election.

19. All persons owning mining ground in this district shall be entitled to a vote.

20. A miners meeting may be called at any time within this district by posting notices in three of the most conspicuous places in the district ten days previous to the time for which the meeting is called signed by the quartz miners of this district stating the object for which the meeting is called and the place designed for holding the same.

21. The above by-laws to remain in force until repealed or altered by a two-third vote at a regular meeting of quartz miners of this district called for that purpose.

22. These laws shall be in force from and after their passage and adoption.

J. O. Robertson, Recorder
Big Bug, Arizona, Oct. 2, 1864

1864.10.17

San Francisco Mining District, Mohave County, October 17, 1864 (From original book of records, pp. 1–3)

The original Records Book of the District was the object of a power struggle as evidenced by the minutes of the meeting of the miners on October 17, 1864. The extent of the district was described in a Report of B. Silliman "On the Mining Districts of Arizona," *American Journal of Science* 41, 2nd ser. (1865), at page 292 as "located upon the eastern side of the Colorado river, Allen's Camp, abbot the center of the district, being stated as 11½ miles from Fort Mojave. Allen's Camp is situated upon Silver Creek so called, a dry arroya which divides the San Francisco District into two nearly equal parts. Measured upon the course of the river, this District extends about twenty miles, or ten on each side of Silver Creek, from north to south. In the other direction, from east to west, the District extends about ten miles, the eastern limit being the first range of mountains, of which the most conspicuous point is known as Boundary Peak."

Meeting of October 17, 1864

At a meeting of the Claim owners of the San Francisco Mining District Territory of Arizona held pursuant to legal notice at Silver Creek in aforesaid Territory Oct 17th 1864.

On motion D. T. McCall was called to the chair and Wm France appointed Sec. Mr Russell stated the object of the meeting as being to enquire as to whether the Office of Recorder in the District was considered vacant on account of the absence of the Recorder and his Deputy from the district and consequent non-fulfillment of the duties pertaining thereto.

After some discussions it was moved and carried that the office of Recorder was now vacant and that it was lawful and proper to elect a Recorder to the vacant office.

Therefore Mr. A. E. Davis nominated C. W. Rowell as canidate for Recorder

M. Welly nominated Jas H. Nicholson as canidate seconded.

Moved and carried that vote by ballot be adopted.

On motion Messrs Morse and Vales were elected as receivers and tellers of votes. Resolved that the polls be kept open for fifteen minutes. On motion the polls were declared closed.

The tellers then proceeded to collect the votes the result of which was as follows.

All of votes for C. W. C. Roswell fourteen no. of votes for Jas H. Nicholson six, total twenty. C. W. C. Roswell was ten declared duly elected.

Resolved and carried that the Records of the District be laid before the meeting for convenience of __ and more especially to ascertain the state of election of Recorder.

Resolved that Messrs Davis, Thompson and France be appointed a committee to demand the Books from the custodian. The subject was as follows: that the date of election was 28th March 1864 and that the custodian (Reddick) refused to deliver up the Books to any person but Col Aikens Depty-Recorder or his order.

On motion it was resolved that the President appoint a Committee to demand formally the Books and papers pertaining to the Recorders Office from the person now holding them. The Pres't. appointed the old committee. The committee acted upon and reported as follows.

That the custodian (Reddick) refused "in toto" to deliver the Books without an entry from the Depty-Recorder Col Akins. Moved and carried that the meeting be resolved into a committee of the whole to obtain the Books "nolens volens" the same being forcibly and most illegally detained. After some discussion the above motion was reconsidered.

Resolved that it be right and proper for the Recorder C. W. C. Roswell to purchase a new set of Record Books pending the absence of the late Depty-Recorder.

Resolved that a committee be appointed to wait on the ex Dept'y-Recorder on his election the District and hand him a copy of the minutes of the above meeting and demand from him the Books of Records of the district.

On motion the President was authorized to appoint a Committee for the above purpose.

The Pres committee was as follows: E. E. Clough, Wm France and R. A. Rise.

Moved and carried that the above proceedings be in force from this date. Moved and carried that the meeting be adjourned "sine die."

Meeting declared adj
L. F. McCall Pres
Wm France Sect

1865.01.11

Agua Frio (Agua Fria) Mining District, January 11, 1865 (Shown on 1866 GLO Report, Rand, McNally, USGS Report, and ABM Map, King, 265)

Laws of the Agua Frio Mining District

At a meeting of the undersigned citizen miners held Jan. 11th A.D. 1865 at the point on the Agua Frio known as the Beaver Dams, the following rules & regulations for governing this Mining District were adopted.

Sec. 1st This District shall be known as the Agua Frio District.

Sec. 2nd The boundaries of this District shall be as follows—On the North by Bradshaw, Turkey Creek and Walnut Grove Districts. Thence running southward along the eastern line of the Wickenburg District fourty miles, Thence east ward fourty miles crossing the Agua Frio at or near the Frog Tanks, thence north fourty miles, Thence westward, to the south east corner of the Bradshaw district.

Sec. 3rd There shall be one recorder elected for this District who shall hold his office one year from the present date, whose duty it shall be to keep a faithful record of all claims of mineral or auxiliary lands and other documents filed with him for record.

Sec. 4th The recorder shall receive for his services one dollar for each claim recorded, & one dollar for each certificate of record.

Sec. 5th All claims shall be numbered each way from the discovery claim.

Sec. 6th The Territorial law for the registry & government of minerals & mineral deposits is hereby adopted, to its full extent and effect.

Sec. 7th These laws shall remain in force until changed by a meeting called for that purpose consisting of not less than twenty interested miners.

Sec. 8th It shall be the duty of the recorder to call a meeting when required by seven or more interested miners of the District and no meeting shall be considered legal unless called by the Recorder who shall give due notice thereof.

Sec. 9th On motion Milton Hadley was unanimously elected Recorder of the District.

Milton Hadley, N. L. Griffin, Alfred Zimmerman
N. P. Pierce, W. R. Basham, W. F. Banning
G. W. Smith, Jesus Munguis, Joseph Lennon
G. Johnson, Simon Roblez
George Clinton, Jesus Otaro
Milton Hadley Secretary, Joseph Lennon President

1865.02.07

Big Bug Mining District, Yavapai County, February 7, 1865 (Published in *Arizona Miner*, July 25, 1865, p. 1:2, and King, 266, shown on Rand, McNally, 1866 glo Report, usgs Report, and abm Map)

In pursuance to a notice a miners meeting was held at Boggs & Companys camp on Big Bug Creek in the Woolsey Mining district county of Yavapai Territory of Arizona Feb. 7th 1865. There being a large majority of the miners of the district present the following business was transacted.

On motion Jackson McCracken was duly installed into the chair and J. M Boggs elected secretary.

The following resolutions were introduced & unanimously adopted as laws regulating the mining interests of the district.

Resolved, That the name of this district shall be known hereafter as the Big Bug Mining district.

Resolved, That a new Recorder shall be elected for this district.

Resolved, That the Recorder be allowed the sum of 50 cents a mile, where the distance exceeds two miles, going to a mine or mines; for recording and returning from a mine the recorder shall receive no compensation for milage.

Resolved, That the recorder be instructed to number all claims located in this district from the discovery claim.

Resolved, That all claims located in this mining district shall not exceed two hundred feet on any Quartz ledge, vein or Lode excepting the discoverer or discoverers, who alone shall be entitled to locate one claim by preemption and one for discovery.

Resolved, That the mining law passed by the Legislative assembly of the Territory of Arizona, concerning the locating of auxiliary land for mining & mill purposes, be adopted.

Resolved, That all claims on Quartz ledges, Veins or Lodes, taken up or located and recorded prior to the date of this meeting, in the name or names of females in this district, shall be considered legal and valid and their rights to said claim or clams shall be respect in the same manner and form that the laws of this district extend to males.

Resolved, That the recorder be instructed to hold his office in the most populous section of this mining district.

Resolved, That all laws passed and adopted by previous miners' meetings held in this district shall be declared null & void and of no effect whatever, such laws conflicting with acts and resolutions passed and adopted by this meeting.

Resolved, That any claim or claims located in this district, shall be placed on record within thirty days after discovery, or such claim or claims shall be liable & subject to re-location by other parties.

Resolved, That whenever the recorder be required by five or more miners of this district to call a miners meeting he shall proceed to have three public notices posted in the most conspicuous places in the district, and specify the motive & object for which the meeting shall be called for.

Resolved, That the proceedings of this meeting shall be published in the "Arizona Miner."

John H. Marion was nominated and elected by acclamation Recorder of the district.

The following was offered and unanimously adopted:

Resolved, that the newly elected Recorder Mr. John H. Marion, be instructed to pursue a legal course in securing all books records or paper belong to this district.

On motion the meeting adjourned sine die—

J.M. Boggs Secretary.
Jackson McCracken President
Big Bug & Feb. 7th 1865

1865.11.11

Wauba Yuma Mining District, Mohave County, November 11, 1865 (From original recorder's book)

This is to give notice that a meeting of the miners of the Wauba Youma mining District Mohave County Arizona territory will be held at the Little meadows in the Wauba Youma District for the purpose of electing a Recorder and transacting any other business that may come before the meeting on the 11 day of Nov. 1865 at 12 oclock M.

Wauba Yuma District, Nov. 11th 1865

Robert A. Rose	F. M. Jackson
Daniel Smith	D. T. McCall
E. M. McCarthy	J. H. Bateman
Sam Knodle	D. O'Leary
N. H. Vosler	Jos Coulter

The meeting convened as per appointment. Mr. Rose called the meeting to order and read the notice as the Laws of the District direct and move that N. H. Vosler be President of the meeting, seconded & carried.

Mr. Rose was nominated as Secretary of the meeting, Carried.

The President then stated the purpose of the meeting briefly

1st the office of recorder being vacant as his time run out on the second day of November it was necessary that a Recorder should be Elected to fill the vacancy.

Mr. D. T. McCall was nominated as Recorder for the ensuing term of the Wauba Yuma Mining District, as no other nominee was proposed, Mr. McCall was declared duly Elected.

Mr. Rose then explained the disadvantage that the miners had to contend with and read the following Resolution which he submitted before the meeting which after due consideration was unanimously adopted.

On account of the scarcity of provisions and mining supplies generally and distance from any white settlement and consequently the many disadvantages to encounter.

Therefore be it Resolved that the Wauba Yuma mining District be laid over until the first day of June A.D. 1866 and all mines that are not represented within one month after the expiration of said time shall be considered forfeited and liable to relocation.

Mr. Smith then stated his views concerning the Territorial mining law and offered the following Resolution which was contested and finally carried

On account of the mining law of the Territory allowing the mines thereof to be held without work on account of Indian difficulties we consider it detrimental to the interest of Mojave County Therefore be it

Resolved that the miners of Wauba Yuma District request their Representative when assembled in the Territorial Legislature to use the influence in exempting Mojave County from said act.

Mr. McCall then moved that the Secretary shall forward a copy of the foregoing Resolution to our Representative of Mojave County. No other business being presented before the meeting it was moved and carried that the meeting be adjourned sine die.

R. A. Rose secretary, H. Vosler pres
Little Meadows Wauba Yuma Mining District,
Moj County, Arz. Ter. Nov. 11th 1865

Other Entries

Claim notices followed, and a notice dated April 20th, 1866 claimed 4,000 feet as a mill site. The final notice in the Book read:

Hardyville, Feb. 24th 1867

This Record Book was found among the effects of D. T. McCall, deceased who was murdered by the Indians on Thursday morning the 21st day of February 1867 in the pass known as "Union."

James P. Bull
County Recorder

1866.05.30

Beaver Mining District, Yavapai County, May 30, 1866 (Published in the *Arizona Miner*, June 13, 1866, 5:3)

On Wednesday, May, 30, 1866, the party held a meeting, formed a new district which they named the "Beaver Quartz Mining District" elected Robert J. Osborn, Recorder, and adopted the Territorial Mining Law as the law that shall in all cases govern persons locating or working mines in said district. Beaver District is bounded as follows:

Commencing on the Agua Frio river, at the mouth of Coyote creek and following up the Agua Frio, north, to the mouth of the Big Bug creek; thence west to the head of Rabbit creek; thence south to the head of Coyote creek; thence east, down said creek, to the Agua Frio.

1870.01.31

Pine Grove Mining District, Yavapai County, January 31, 1870 (Published in the *Weekly Arizona Miner* February 25, 1871, 1:1)

At a regular organized meeting of the miners of "Pine Grove District," held in said district on the 31st day of January, 1870, the following laws and resolutions were adopted:

Section 1st. This district shall be known as the "Pine Grove Mining District," and shall include all that section of country known as the "Bradshaw Mountains," not included in any other lawfully formed mining district.

Sec. 2d. That all claims located on quartz lodes, in this district, shall consist of two hundred feet following the main lode with fifty feet on each side of said lode, allowing, in all cases, two hundred feet extra to the discoverers.

SEC. 3d. That no work shall be required to be done, in order to hold any claims in this district until the cessation of Indian hostilities.

SEC. 4. That the Recorder shall be allowed one dollar for each notice of one name, and one dollar and fifty cents if more than one name; and in all cases, if so required, shall give to the claimant a certificate, showing the time the record was made.

SEC. 5th. That all placer claims shall not exceed three hundred feet in length and one hundred feet in width.

On motion, Joe Melchoir was elected Recorder, after which the meeting adjourned until further notice.

1870.03.08

Martinez Mining District, Yavapai County, March 8, 1870 (Ledger containing location notices found in Sharlot Hall Museum Archives)

District organizational documents not found, but location notices exist in a Martinez Mining District Recorder's book at the Sharlot Hall Museum Archives. Willard Rice, apparently the first recorder, and Jas. A. Farnsworth, deputy recorder. Later early notices recite George H. Kimble, as recorder.

1870.10.24

Bradshaw Mountain Mining District, Yavapai County, October 24, 1870 (Shown on USGS Report, and regulations published in King, 267)

In pursuance to notice a meeting of miners of Bradshaw mountain was held Oct 24th 1870 at the mining camp of Messrs. Taylor McCracken & Co for the purpose of organizing a mining District & enacting laws for the Government of the same—

On assembling M Taylor was elected chairman who explained the object of the meeting & appointed McMorris, Geo Monroe & Samuel Boll a com of three to draft a code of mining laws for the Dist, describe its boundaries & give it a name—

After a short absence the com presented to the meeting the following report—

ART 1 That this Dist be known as Bradshaw Mt Dis't & bounded as follows, on the north by Turkey Creek Mining Dist, on the west by Bradshaw Dist on the south by Pine Grove Dist & on the east by Black Canon Dis't—

ART 2 That this Dist elect a recorder whose term of office shall be one year from date of election & whose duties shall be to make true & correct entries of notices of mining & all other claims left with him for record in good & suitable books expressly for that purpose, said books of record to always be kept in the possession of the Recorder & subject to inspection to all persons when calling at proper office hours.

ART 3 The Recorder shall receive as compensation for his services the following fees viz—Recording mining & all other claims one dollar for each name & will be required to give a certificate of record to all persons requiring the same, by paying for each certificate one dollar—

ART 4—Any six miners by giving three days notice regularly posted at conspicuous places at different mining camps in the Dis't, can call a regular miners meeting & it will be the duty of the Recorder to act as Secretary of all regularly called miners meetings—

ART 5—No person will be allowed to vote at miners meetings except those owning interest in the Dis't—

ART 6—A claim of any Gold, Silver, Copper, Lead or other metallic quartz or rock on any lead, lode, or ledge, or deposit shall be two hundred feet by pre emption & two hundred feet additional for the right of discovery together with all dips, spurs, angles & variations, and seventy-five feet on each side of the Leade, lode or ledge so claimed, for working & mining purposes—

ART 7—Any person or persons discovering any Leade, lode ledge or deposit of Gold, Silver, Copper Lead or Clay & claiming the same will be required to erect a monument on the Leade with notice naming the leade stating the number of feet claimed describing as near as possible its general course & locality—

ART 8—Any person or persons locating any Leade, lode or ledge or deposit of Gold Silver, Copper, Lead or other metallic bearing quartz rock or clay will be required to file a notice with the Recorder for record within ten days after such location or locations—

ART 09—A monument upon any leade, lode ledge or deposit will be sufficient to hold an individual or company's claim six months from the date of record—

ART 10—A shaft or Tunnel of ten feet on the leade lode or ledge of any individual or company's claim will be sufficient to hold the same one year from the date of record & a shaft or tunnel of twenty feet on the leade lode or ledge on any individual claim or company's will be considered a permanent title & not subject to relocation—

ART 11—Any individual or company not worked for one year from the date of record will be declared abandoned & subject to relocation—

ART 12—Any reasonable amount of ground not exceeding ten acres & taken in a body can be located for a mill site for milling & mining purposes together with wood timber & water on the same & must be recorded giving its proper boundaries

ART 13—No person or company can nor shall in any way obstruct the free access to or from any individual or company claim (mining) nor infringe in any way by cutting wood or timber on ground allowed each side of leade for working & mining purposes—

ART 14—Water claims for milling and mining purposes can be located upon any gulch ravine or spring including & claiming the entire amount of water running in such gulch or ravine to the point of location & amount of water afforded by such spring if actually necessary for milling & mining purposes by the person or company so locating & no person or company can in any manner make subsequent locations of water privileges that will in any way affect or diminish water previously located. Water privileges must be recorded with description of their locality & boundaries of claims—

ART 15—A lot on any town site in this Dist shall be one hundred feet square & subject to preemption & must be recorded—

ART 16—Grounds for agriculture or garden purposes can be located by filing notice with the Recorder giving the boundaries & location provided such location does not in any manner interfere with the free access to or from any mining claims or interfere with or diminish any water claimed & required for milling & mining purposes—

ART 17—The rights of miners, for milling, mining & working purposes shall have preference over all other rights.

MACK MORRIS
GEO MONROE
SAMUEL BALL
Committee

On motion the report was received & the committee discharged—
On motion Mack Morris was elected recorder for the Dis't.
No more business being before the meeting it was adjourned sine die
MCMORRIS—Sect'y.

1870.11.04

WALLAPAI (HUALAPAI) MINING DISTRICT, MOHAVE COUNTY, NOVEMBER 4, 1870 (FROM KING, AS PUBLISHED IN MOHAVE COUNTY *WEEKLY MINER* ON DECEMBER 2, 1871 AT 4:1, 2, AND RECORDS OF THE DISTRICT FROM BLM RECORDS)

Notice Is hereby given that there will be a meeting of the claim owns at Soldier Springs, in the Sacramento Mining District, on Friday, November 4th 1870, for the purpose of altering and making new laws to govern said district and changing the name of said District and to do such other business as may be brought before the meeting

James Fleming
Samuel Todd
H. E. Davis
Geo Okay
Berry H. Spear
Wm Fee
R. C. Todd
James P. Bull
Geo Fisher
M. E. Doolittle
Dan H. Smith

Soldier Springs, Nov 4th 1870

In pursuance of the above published notice (which was in accordance with the old laws of the District) a meeting of the claim holders assembled at Soldier Springs on the 4th day of November 1870 & on motion of Mr. O Key, James Fleming was chosen President & Henry Hardy Secretary.

On motion of Mr Dunkel a committee of three consisting of Messrs Hardy, Hoffner & Fleming were chosen to draft a new set of laws to govern the Dist on & after this date—

On motion all laws, rules & regulations heretofore adopted are hereby repealed & declared null & void—

The committee appointed to draft a new set of laws to govern the Dist reported the following which upon motion of Mr Fee were adopted to wit.—

Art 1st This District shall be known hereafter as the "Wallapai Mining Dist" & shall be bounded as follows to wit—"Commencing at Beals' Springs & running thence along the Prescott & Mohave toll road westerly ten miles, thence up the Sacramento Valley centrally with the river range of mountains thirty miles, thence East twenty miles, thence down the Wallapai Valley to road & thence to Beal's Springs"—

Art 2d Each locator of a claim upon any mineral bearing vein or lode shall be entitled to 200 feet in length & one hundred & fifty feet on each side, if not conflicting with any other parallel leade or lode & each discoverer of any vein or lode shall be entitled to 200 feet additional—

Art 3—All locations shall be plainly marked by stakes or monuments & notices posted defining the boundaries & containing the name of the claimants & number of feet claimed—

Art 4th All claims located from the first of October 1870 to Nov 4th 1870 shall be duly recorded by the 15th of November 1870 & all claims located after the 4th day of Nov 1870 shall be recorded within ten days after location & as soon as the notice is filed it shall be considered recorded—

Art 5th It shall be the duty of the Recorder to keep a suitable book or books a full record of the proceedings of all meetings free of charge, and to place on record all locations brought to him for record, where such location does not interfere with or affect the right or interests of others, provided he shall receive the sum of fifty cents for each location of 200 feet—

Art 6th The Recorder shall not recognize any location of any character made prior to the first day of Oct 1870—

Art 7th The Recorder shall be required to visit every location made, prior to recording the same & see that the boundaries are well defined by good & substantial marks & shall receive there fore the sum of fifty cents per mile to be computed from O Key or Whitney Springs—

Art 8th In case the Recorder neglect to perform any act required of him by these laws he shall forfeit his office, provided that three notices be posted calling a meeting giving at least two days notice signed by three claim owners—at said meeting the parties aggrieved shall submit their charges & if sustained by a majority of those present the office shall be declared vacant & his successor shall be immediately elected—

Art 9th each & every claim holder shall be required to put at least two days work upon each location for every two hundred feet claimed within ninety day's after recording—And after the expiration of said ninety days each & every claim holder shall be required to put at least one day's work upon each location for every two hundred feet claimed every thirty days, until the depth of thirty feet is sunk on the location—after the depth of thirty feet is sunk it shall not be subject to re-location, unless publically abandoned, for one year—

Art 10th Every claim holder, locator or locators not complying with the provision of article nine shall forfeit all rights of every kind, nature & character whatsoever & the locations made by them shall be considered & treated as null & void—

Art 11th Two or more companys owning claims upon the same vein or lode may join together in prospecting said vein or lode & and the work done on the vein or lode by one company may be accredited to both, provided articles of agreement to that effect are filed with the recorder & one of the locators make oath that it is a bona fide transaction—

Art 12th The officers of the Dist shall consist of a Recorder to be elected by those taking part in these proceeding & who shall hold his office for the term of two years from the date of this election & may have one or more deputies—

Art 13th It shall be the duty of the Recorder to give any claim holder, who shall have a claim recorded in his books a certificate the same & for such services shall receive the sum of one dollar & fifty cents—

Art 14th It shall be the duty of the Recorder to give locators & owners of claims who shall make a written affidavit that he has done a certain amount of work on a vein or lode & receipt for the same & for said services he shall receive the sum of one dollar & fifty cents & said amount shall include the expense of filing the affidavit—

Art 15th These laws shall be subject to revision alteration or amendment by a majority of claim owners & actual residents of the Dist at all time, provided, that twenty-five claim holder cause to be posted in three public places (including one at the county seat) notices setting forth the place, and intended revision, alternation or amendment to be made—

Art 16th The Recorder shall be a bona fide resident of the Dist & shall keep all books & papers in said Dist, so long as it may be safe to do so, & in case of removal shall deposit all books & papers with the county Recorder subject to the order of the claim holders of said Dist, provided that in the case of removal of said books & papers from the office of the Count Recorder said claim owners give sufficient guarantee for their safety—

Art 17th The Recorder shall not absent himself from the Dist without having a good & reliable deputy & in case he should be absent for more than thirty days said office shall become vacant & said deputy shall immediately call a meeting of claim holders whose duty it shall be to elect a Recorder to fill the vacancy. In case there shall be no deputy then three (3) claim holders shall be empowered to call a meeting for a like purpose & either of said notices shall give at least five days notice & one copy shall be posed on the county seat—

Art 18th There shall be held at the office of the Dist Recorder an annual meeting of claim owners at the hour of 10 o'clock a.m., the first Wednesday of November

On motion of Mr Speer, William Fer was duly elected Recorder for the ensuing two years

On motion the meeting adjourned sine die—

Henry F. Hardy
Secretary

James Fleming
President

Meeting of March 15, 1871, at Cerbat City

In pursuant of the above all of the miners of Wallapai District, A meeting was held at Cerbat City on the 15th day of March 1871 for the purpose of amending the By Laws of said District.

On motion S. M. Atchison was called to the chair and J. Slarer was appointed Secretary.

On motion of Geo Okay the meeting was adjourned to meet on Wednesday the 22nd day of March 1871

Meeting of March 22, 1871, at Cerbat City

Miners meeting called to order & S. M. Atchison chosen chairman when the following amendment were made to the mining laws of the Dist—

Article was so amended to read.

That two days work in each company location of two hundred feet or more will be sufficient to hold their claim for the term of three months—

Article 17 Is thereby amended so as to read as follows, "The Recorder shall not absent himself from the District without having a good & reliable Deputy" striking out the words thirty days

Article 19. Whereas we think it to the interests & advancement of this District to those having mines that they do work on the furnace shall have their work or such amount thereof as is necessary to hold their claim accredited to such claims for the term of three months—

Article 20—The Recorder shall be required to enter upon the records the appointment of his deputy or deputies—

Art 2 Is so amended as to confirm to Senator Stewart's bill, which is an act to protect the rights of miners, a width of one hundred & fifty feet on each side of the thread of the lode, shall be allowed, and all the veins found within the space formed by this surface location extended downward perpendicularly shall belong to the locator of the claim with the right to follow such lodes even if they dip out of said lines

Meeting then adjourned

W. J. Fee Recorder—
S. M. Atchison—President

Meeting of November 4, 1872

Meeting held Nov 4th 1872 at which Mr Toll moved that the congressional act now in force be adopted as a whole, which was adopted—

Dr Rees moved that twenty days work be considered equivalent to $100.00 done in a ledge, carried—

A. C. Haskell receiving 62 votes out of 94 cast was declared elected Recorder

H. F. Baker Presdt—

1871.05.13

Tiger Mining District, Yavapai County, May 13, 1871 (Published in *Weekly Arizona Miner*, May 13, 1871, 3:2)

Article I. That the District is to be known as the "Tiger District."

II. This district shall include all the country draining into a stream known as the "Humbug" from its bend to a point ten miles down said stream.

III. The officers of the district shall consist of one Recorder who shall be elected by the miners of the District for the term of one year, unless sooner removed by charges properly preferred and sustained, by claim holders of the district.

IV. It shall be the duty of the Recorder, when so ordered by five claim holders to call and preside over all meetings upon business pertaining to the district.

V. The duties of the Recorder shall be to record all claims and articles pertaining to the rights of the miners. Also, to act as Secretary of all meetings, keeping a correct record of the proceedings of the same.

VI. All locations made in this district shall be designated by a monument and notice thereon to be placed on some prominent part of the location describing location and direction of the lode, and the amount of ground claimed not to exceed 200 feet to each claimant, with the right of an additional 200 feet for the discovery claim.

VII. It shall be the duty of every locator to file a copy of his notice, within ten days from date of location, with the Mining Recorder of the District.

VIII. Resolved, That five days work to every claim shall hold the same for a period of six months, the work to be commenced within thirty days from date of record.

IX. The Recorder shall be allowed one dollar per mine, as fees for recording.

X. On motion of J. Reese, E. A. Gordon was duly elected Recorder for the period of one year.

1871.11.20

Sedgwick Mining District, Yavapai County, November 20, 1871 (Published in King and on January 6, 1872 in *Weekly Arizona Miner*, 3:2)

Mr. Pierce Dorgan was duly elected chairman & forthwith proceeded to call the meeting to order—C. G. Terry elected chairman—

1—On motion of Mr. Cox a District was formed having as a pivot a high point to the North east of Clinton station, said point to be hereafter known as McCloud's point—

2—On motion the following described lines is hereby made the boundary of the Dist now undergoing organization—From McClouds point east for a distance of eight miles & from thence south a distance of 12 miles thence west 1 miles, thence north 12 miles thence 4 miles east to McClouds point comprising a section of county 12 miles square—

3—On motion Dist name Sedgwick Dist.

4—Regarding claims by motion of Mr. Hall that a period not to exceed twenty days after location be granted to claimants to record their respective claims before the Dist recorder—Carried—

5—Motion of Mr. Kennedy that five days work on any claim or company's claim hold the same for six months—Carried—

6—Motion that within 30 days after recording work be commenced upon said recorded claims—carried—

7—Motion Mr Cox that each claim or company's claim have their respective limits defined by erecting a monument or driving a stake at the beginning and terminus of said claims—carried—

Motion of Chairman, Samuel Hill was elected Dist Recorder in & for Sedgwick Dist.

Motion of Mr Kennedy that the Dist recorder have & receive for a consideration the sum of one dollar for each claim placed upon record—

Resolved that a written application be presented to the Recorder of (said) Sedgwick Dist when a miners meeting is to be held & subscribed to by twelve miners interested in said District & be it,

Resolved that the Recorder give ten days notice by placing or having placed in three conspicuous places in said Dist before such or any meeting can be held pertaining to the interests thereof—& be it further

Resolved that the Recorder be required to keep a book for recording claims & said book be open to public inspection & the Recorder will cause to be placed upon the first pages of his record book the proceeding of this meeting & be it further

Resolved that the aforesaid laws are in force from & after this date & all laws conflicting with the same are hereby repealed.

Pierce Dorgan, Chairman
C. G. Terry Secty
Samuel Hill, Recorder

[*Note*: The published version is much more straightforward, but substantively the same.]

1872.01.27

Fort Rock Mining District, Yavapai County, January 27, 1872 (Published on February 17, 1872, in *Weekly Arizona Miner* at 3:3)

Laws of Fort Rock Mining District

At a meeting of miners held at Fort Rock, in the county of Yavapai, Territory of Arizona, on the 27th day of January, A.D. 1872, for the purpose of organizing a mining district, E. C. Bradshaw was elected Chairman and Robert Kelly was appointed secretary of the meeting.

On motion of S. M. Ball, the following code of By-laws to govern the District was unanimouslly adopted, to wit:

Article 1. The District shall be known as the Fort Rock Mining District, and is bounded as follows: Commencing at a point twenty miles due east of the Fort Rock house and running due North for a distance of twenty miles, thence due west for a distance of forty miles, thence due south for a distance of forty miles, thence due east for a distance of forty miles, thence due north for a distance of twenty miles or to the place of beginning.

Art. 2d. A locator of a ledge shall be entitled to hold 200 feet in length and 100 feet in width for working purposes, together with all dips, spurs, angle and offshoots belonging to the ledge located; and the discoverer of a ledge shall be entitled to 200 feet extra, but no person will be allowed to hold more than one claim by location on any ledge in the district.

Art. 3d. All locations made must have a written notice, posted on a monument not less than two feet in height; the notice shall state the number of feet claimed, also the names of the locators and the locality of said claim relative to some nature or artificial object or well established lode.

Art. 4th. After a notice has been posted on a claim, ten days shall be allowed to have the same recorded. A failure to do so shall subject the claim to re-location.

Art. 5th. After a claim has been recorded, thirty days shall be allowed to do the assessment work thereon, which shall consist of removing or excavating earth, rock or ledge matter to the amount of forty cubic feet, which shall hold the claim for the space of one year from its date of record. A failure to comply with the conditions of this Article shall subject the claim to re-location.

Art. 6th. The Recorder shall make all records in a book kept for that purpose, and shall deliver the said books to his successor when elected, and shall receive for his fee the sum of fifty cents for each location of 200 feet.

Art. 7th. The recorder shall have power to appoint one or more deputies; and all examination of the records shall be made in the presence of the Recorder or his deputy.

Art. 8th. A recorder shall be elected on the 27th day of January of each year, and shall immediately thereafter assume the duties of his office, which he shall hold for the space of one year or until his successor is elected.

Art. 9th. Whenever an locator or company shall have sunk a shaft to the depth of 20 feet on their claim, the ground claimed shall be considered as belonging in fee to the locator or company, and shall not be subject thereafter to re-location.

Art. 10. These laws shall not be altered or amended for the space of one year.

On motion of Wm. Burch, Robert Kelly was elected Recorder.

On motion, the meeting adjourned sine die.

E. C. Bradshaw, Chairman
Robert Kelly Secretary

1872.02.20

Copper Mountain Mining District, Yavapai County, February 20, 1872 (From BLM records and as recorded in Graham County Book 1 of Misc. Records, pp. 1, 56–57 and 128–29)

February 20th 1872

At a meeting of Miners, held this day at Copper Mountain, Arizona Territory the following proceedings were had:

James Ballard was chosen president and John W. Reed Secretary the object of the meeting was declared to be to propose and adopt a code of mining laws and define the boundaries of the Mining district.

N. L. Rynerson,[1] Joseph Yankie and I. H. Itone were appointed a committee to prepare a code of laws for the district. It was Resolved that the limits of this District is to be known as Copper Mountain Mining District Shall be as follows it wit:

On the East the San Francisco River, on the South the Gila River, on the west Eagle Creek and on the North a line drawn from Francisco River Westerly to Eagle Creek passing fifteen miles north of Copper Mountain.

Jas B. Bullard, President
John N. Reed, Secty

[*Note*: Here follow the rules of the district that are wholly inconsistent with referring only to compensation of recorder and manner of location, which are afterwards changed and made to conform with mining laws of US of May 10, 1872.]

Meeting of September 10, 1872

At a meeting of the Miners held at Copper Mountain Mining District there were present James Ballard, J. S. Yankie T. Stevens, A. Sibel, Captain M. [Miles] Joy,[2] J. W. Reed, J. Colwell, Owen Roberts, E. M. Pearce and others for the purpose of Amending the district laws.

On motion of J. N. Reed was elected President and E. M. Pearce Secretary. It was recorded that the Mining Laws of May 10th 1872 enacted by Congress be hereafter the law of this district so far as applicable.

Resolved, That until a larger force of miners can be introduced here, in order that they may be able to maintain themselves with safety against hostile Indians, the records of this mining district be kept at Silver City New Mexico,

Resolved that a claim may be fifteen (1500) hundred feet in length and six (600) hundred feet in width.

Resolved that a claim shall be recorded within thirty days from date of location.

On motion adjourned.

J. W. Reed President
E. M. Pearce, Secy
Copper Mountain district
Yavapai County, A.T.
September 10th 1872
James Bullard, Recorder

I hereby certify that this a true copy of said meeting as appears in Record Book of Copper Mountain Mining District, Yavapai County, A.T.

E. S. Goulding, Recorder

Meeting of October 17, 1873

Clifton, Arizona Territory

Yavapai Co. Oct 17th 73
At a meeting of the miners of Copper Mountain District which convened at the above mentioned place and time, the following is a record of their proceedings.

The meeting was called to order by R. B. Metcalfe [Robert Baylor][3] when H. M. D. Hewett Esq. was elected as Chairman, Edwin Polk as secretary and Eugene S. Goulding as recorder for the district during the ensuing twelve months.

It was resolved that the mining laws enacted by Congress May 10th 1872 should still be recognized as the laws of said district.

Resolved that the records of this mining district shall hereafter be kept at Clifton, Yavapai Co. Arizona Ty

There being no further business brought before the house, on motion the meeting adjourned.

H. M. D. Hewett, Esq., Chairman
Edwin Polk, Secy
Eugene S. Goulding, Recorder

I certify that the enclosed is a true copy of said meetings as it appears in Record Book of Copper Mountain Mining District Yavapai County, A.T.

E.S. Goulding, Recorder
Clifton, January 26th 75

Meeting April 6, 1881 (From District Records and Graham County records Book 1 of Misc. Records, p. 1)

Whereas at a meeting regularly held in Copper Mountain Mining District and in Greenlee Gold Mountain District the undersigned were duly appointed a committee to adjust the boundary line between the said Mining districts. Therefore after mature deliberations we the committee do Resolve

1st That the said boundary line is thereby established and shall be as follows viz: From a point on the Frisco River known as Lezinskys Dam in a direct line Westerly to the stone cabin on Chases Creek thence Northerly following the bed of said Chases Creek to its source, thence continuing Northerly to Walnut Spring.

2nd that all records theretofore made in either District for locations made on Territory hereby affected shall be considered valid and remain in full force.

3rd That a copy of these resolutions be furnished to the Grant County Herald for publication, also to the recorders of each of the district interested also to the Recorder of Graham County Arizona for record. Dated at Clifton April 6th 1881.

By the committee

Wm Bonnell	E. B. French
Joseph F. Yankie	Sam J. Freudenthal[4]
For Greenlee Gold Mountain District	For Copper Mountain Mining Dist.

Recorded at the request of S. J. Freudenthal 16th day of April 1881 at 30 past three oclock P.M.

Wm F. Clarke Recorder
By Wm Osborn Deputy

Office of the recorder of the County of Graham Territory of Arizona. Know all men by these presents that I Wm F. Clarke Recorder of the County of Graham territory of Arizona do hereby appoint Wm Osborn Deputy Recorder of this the County aforesaid to act in such capacity during my pleasure;

In witness whereof I have hereunto set my hand and affixed the seal of my office this 16th day of April 1881.

Wm F. Clark, Recorder

Meeting of January 2, 1882 (Recorded in Book 1 of Misc. Records, pp. 56–57)

Samual J. Freudenthal elected Recorder. Motion:

Resolved That Five dollars per day be allowed in performing assessment work on all claims located within the Copper Mountain Mining District for each man actually employed in performing such assessment work.

Meeting of December 31, 1883 (Recorded in Book 1 of Misc. Records, pp. 128–29)

Sig Heisl elected Recorder for 1884.

The following resolution was then presented by Fenton Ingraham and in motion duly put and carried was unanimously adopted as one of the Bylaws of the District. Resolved: that fourteen (14) days work on each claim be allowed as performing assessment work on each claim in this the aforesaid District.

1872.09.10

CEDAR VALLEY MINING DISTRICT, MOHAVE COUNTY, SEPTEMBER 10, 1872 (RECORDED IN BOOK 1 OF MISC. RECORDS, PP. 182–84, AND PUBLISHED IN *WEEKLY ARIZONA MINER*, DECEMBER 14, 1872, 4:1 AND IN GLO MICROFILM, SHOWN ON RAND, MCNALLY, 1863 GLO INDEX, USGS REPORT, AND ABM MAP).

Mining Laws of the Cedar Valley District

At a meeting of the miners at the Gunsight mine, situate on the eastern slope of the Wallapai range of mountains, about forty (40) miles south of Wallapai spring and about twenty-five (25) miles a little north of east from Needles, Mohave county, Arizona Territory, held September 10th, 1872, for the purpose of forming a new district and to make laws for its government, the following resolutions were adopted, to wit:

Resolution 1st. This district shall be known as Cedar Valley District, and shall be bounded as follows: Commencing at the lower Indian gardens on the Big Sandy, about five miles below the mouth of Cedar Valley wash, and running thence west to the western base of the Wallapai range, thence northerly with said western base to the south boundary line of the Maynard District, thence with said line East to the Big Sandy, thence down said Big Sandy to the place of commencement.

Resolution 2nd. The Congressional mining law, approved may 10, 1872, and now in force, shall, with some additional rules hereafter prescribed, be the law of this district.

Resolution 3rd. This district shall have a Recorder, who shall be required to reside in the district, and who may hold office for one year from the date of his election, or until his successor shall have been duly elected, when (on the demand of such successor) it shall be his duty to turn over, in good order and without charge, all books and papers appertaining to such office.

Resolution 4th. The Recorder of this district shall be required to keep (at his own expense) suitable books for recording all notices or other papers required to be recorded, and he shall be entitled to receive one dollar for each name on each notice recorded by him, and reasonable rates for all other office work; and such records shall be subject to inspection, without charge, at all time, by any person interested in the district, but such inspection shall not be allowed except in the presence of the Recorder or his deputy.

Resolution 5th. A notice conspicuously placed on the ground claimed thereby, with the boundary monuments established as required by law, shall be sufficient to hold for thirty days from the date thereof but if such notice be not recorded or filed for record in such time, the ground claimed thereby shall be deemed subject to re-location.

Resolution 6th. Sam. A Pearson is hereby elected Recorder of this district.

Andrew B. Peterson, President
Sam. A. Pearson, Secretary.

Amendment of May 15, 1872

Resolution 3rd Is hereby amended to read as follows—This district shall have a recorder whose duty it shall be to record all notices of claims presented for recorded, provided that the recorder or his deputy shall have visited the claim so required to be recorded, and provided further that the recorder shall appoint some suitable person residing near the Hope or Hibermian locations to act as deputy.

Resolution 4th The words in lines 5, 6 and 7 "for each name on each notice shall receive one dollar" shall be changed to read as follows—and he shall be entitled to receive four dollars for each notice located.

Amendment of September 10, 1873

At a meeting of the miners of Cedar Valley district held September 10th 1873, the following resolutions were adopted, to wit:

Resolution 1st. Resolved that in work the assessment (required by act of Congress approved May 10th 1872) on mines seven and one half (7.50) dollars be allowed for each and every days labor performed to be sworn to by the parties having said labor performed.

On motion the meeting proceeded to ballot for recorder for the ensuing year.

Sam A Pearson, received nine votes (the total number cast) he was declared to be duly elected Recorder for the ensuing year.

Sam A. Pearson, Sec'y

I hereby certify that the above is a true, full and correct copy of the laws of Cedar Vally Mining District

Cedar Valley

January 1st 1874

Recorded at the request of Sam A. Pearson, January 5th A.D. 1874 at 10 A.M.

Jas P. Bull

County recorder

By Caldwell Wright, Deputy

[*Note*: The first meeting notice was published in the newspaper; the additional material is from the recorded copy.]

Amendment of February 11, 1883

Pursuant to a public notice duly posted in this Mining District a Miners' Meeting as called for this day for the purpose of Amending the Location Regulations of Customs in regard to Annual Assessment.

The following were present at the Meeting: N W. Clark, Wm Collins, John Long, W. H. Gellis, Sr., S. Devine, G. Darden, Peter Olinger, Juan Espanosa, Santiago Altimara, Juan Feloya, Rich Holt, Louis Hoebel, J. A. DeLund, John Faley and George Myers.

[*Note*: John Long elected chairman.]

The meeting being called to order by the Chairman, on Motion, the District Law of September 10th 1880, in regard to Assessment Work was declared "Repealed."

On Motion a Resolution was passed that on, and after this date, an "Assessment work" shall consist of (20) Twenty days Labor, or at the rate of Five Dollars per Diem, where it is necessary to "Drill" and Blast.

And of Twenty Five (25) days Labor, or at the rate of Four Dollars per diem where the ground only required the Pick and Shovel.

On Motion, it was resolved that on and after this date an Assessment work on a mine shall be considered to have been completed when the Owners, or Representative, of said Mine shall make "Affidavit," that one hundred Dollars worth of Labor has been performed on said Mine. And they shall have said Affidavit "Recorded" with the District Recorder.

On Motion a resolution was passed that the Recorder shall receive a "Fee" of one Dollar for recording such Affidavit.

1871.11.18

Papago Mining District, Pima County, est. November 18, 1871 (Amendment of July 2, 1873, as recorded in Pima County Misc. Records at page 234–26 and published in *The Tucson Daily Citizen*, July 5, 1873, 4:3)

At a Meeting of the Miners and persons interested in the Papago Mining District held at the office of William J. Osborn in the Town of Tucson on the 2nd day of June 1873 at 7 o'clock P.M. pursuant to the following call and notice.

Tucson, A.T., June 23rd 1873
A. Brichta, Esq., County Recorder Pima County
A.T. Ex officio Recorder Papago Mining District.

The undersigned Miners interested in the Papago Mining District Pima County A.T. respectfully ask you to call a meeting of the Miners of the Papago Mining District as provided by the article 4 of the Mining laws of the District.

Samuel Hughes, H. Stevens, R. N. Leatherwood.[5]

Notice is hereby given to all interested miners of the Papago Mining District that there will be a meeting of the mines of said District at the Office of William J. Osborn, in the Town of Tucson, Pima County A.T., on July 2, 1873 at 7 o'clock P.M.

A. Brichta, Recorder

Copies posted on the door of the office of William J. Osborn, one copy at the Village of San Xavier and one copy at the office of the Montezuma Mining Company

The meeting was called to order by A. Brichta. William J. Osborn submitted for the consideration of the meeting the following draft of the Rules and regulations governing the location, manner of recording the amount of work required to hold possession of a mining claim within the Papago Mining District, Pima County A.T. Which being read, and considered and unanimously adopted as follows to wit:

1st All that portion of the county of Pima situate about thirty-five miles in a southwesterly direction from the town of Tucson, in said county of Pima, bounded as follows: Commencing at a monument of stone five miles distant from the mouth of the main shaft of the Montezuma mine, thence north five miles, thence west ten miles, thence south ten miles thence east ten miles, thence north five miles to the place of beginning, shall exist situate and be known as the Papago Mining District.

2nd All the provisions of the act of the Congress of the United States and titled "An Act to provide the development of the mining resources of the United States," approved May 10, 1872, are hereby extended over and declared in force in governing the location, holding and working of mines in the Papago Mining District.

3rd All locations made in pursuance of these rules and regulations shall be recorded in the Record of Mines in the Recorder's office of the county of Pima, within thirty days of the date of location, and County Recorder of said Pima county shall be ex-officio Recorder of Papago Mining District.

4th Upon application of any three miners interested in Papago Mining District by petition to the Recorder, it shall be his duty to call a meeting of the miners of said district, by posting notices for a period of ten days in three conspicuous places, one of which shall be within the limits of Papago Mining District, and said meeting may be held either within the limits of said mining district or in the town of Tucson, as may be stated in the notices to be posted as above provided; provided that these rules and regulations shall not be altered, amended or changed, except at a meeting held as hereinbefore provided, after thirty days notice.

5th The rules and regulations of the Papago Mining District passed and adopted at a meeting held on the 18th day of November, 1871, be and the same are hereby annulled, abrogated and repealed, provided that nothing in this section contained shall be construed to affect any rights acquired thereunder.

(signed)

Miguel Alvarez
Samuel Hughes
A. Lizard
William J Osborn
Thos Hughes

Jose Fontez
R. N. Leatherwood
L. C. Hughes
H. S. Stevens

A. Brichta County Recorder &
Ex Officio Mg Dist Recorder Tully Ochoa & Co.

1874.08.17

Owen Mining District, Mohave County, August 17, 1874 (Recorded September 7, 1874, in Book 2 of Misc. Records, p. 307)

Owen District

Resolutions

At a meeting of the miners of Owen District held at the Porphyry Tank Mohave Co. A.T. on August the 17th 1874 when the following rules and regulations were adopted.

This District shall be known as Owen District and bounded as follows. Commencing at the southeast corner of Aubrey District and stream known as the "Big Sandy" and running down said stream to the north of the "Santa Marie" thence west fifteen miles thence northerly to the southwest corner of Aubrey District thence along the south line of Aubrey District to the place of beginning.

And be it further resolved that no Recorder shall be elected for this District, and all claims located in this District shall be recorded in the county records within sixty days from date of location.

And be it further resolved that these Laws with the laws of Congress and this Territory are deemed sufficient for the government of this District.

On motion the meeting adjourned sine die.

A. F. Owen Chairman
J. McCrackin Secretary

Recorded at the request of R. Gird, Sept 7th A.D. 1874 at 11 A.M.

1875.02.15

Truman Mining District, February 15, 1875, Pima County (Recorded in Book 1 of Misc. Records, p. 296)

Organization of the Truman Mining District.

Rules and Regulations of the Truman Mining District adapted February 15, A.D. 1875 at a meeting held at Kings Ranch, Pima County, Arizona Territory, pursuant to a written notice of said meeting posted at said Ranch February 11th, 1875.

Notice:

There will be a meeting of all who are in favor of forming a mining district to include the Santa Rita and Cayetano Mountains at Kings Ranch near the old Tumacacari Mission Church on the fifteenth at 8 o'clock A.M.

Wm. G. Boyle, J. C. Truman, J. E. Magee, H. C. Hodge (Salero Mine, February 11th, 1875).

February 15th, 1875 meeting called to order at 9 A.M. Present J. C. Truman, H. C. Hodge, Wm. G. Boyle, John E. Magee, A. C. Benedict, John Mansfield, Cornelius Ryan, Henry Mims and Joseph Kings. On motion of Mr. Hodge, Mr. Boyle was elected chairman.

On motion of Mr. Truman, Mr. Magee was elected Secretary. On motion of Mr. Hodge, the chairman appointed Mr. Hodge and Mr. Benedict a committee on Bye laws and Regulations who presented the following which were read by sections and adopted.

Section One. This mining district shall include the Santa Rita, Cayetano and Tubac Mountains and all outlaying spurs and foot hills of the same; the old, so-called Tumacacari Mission Grant and that portion of Territory lying between the Solero Ranch and the Santa Cruz River including southward to the Sonora River called and known as the Stuibaba.

Section Two. The name of this district shall be the Truman Mining District.

Section Three. The annual meeting of voters of this District shall be on the third Monday of February of each year at which meeting an election of Recorder shall be held and such other business be transacted as may be found necessary; ten days notice shall be given of said meeting by the recorder posting three notices in three of the most public places within said District.

Section Four. There shall be elected at this meeting a recorder for the District who shall hold his office for one year from this date and until his successor shall be elected as provided for in Section Three whose duties shall be as follows: First, to file and record in his office all notices of locations made within the bounds of the Mining District, and every three months have the same recorded in the County Recorder's Office for which purpose he shall receive one dollar for each mining claim he shall record and shall also receive from the person who may have a claim recorded by him a sufficient sum to pay for recording the same in the County Recorder's Office.

Second, before recording any mining claim the Recorder shall perform himself by personal examination or otherwise of the location of any mining claim presented for record and shall in no case make of record in favor of one party when another party to his knowledge claims the same location. Third. The Recorder shall call a meeting of the miners of this Mining District upon application of five miners or residents of the District made in writing and shall post or cause to be posted notices of the same for at least thirty days before the meeting in three of the most public places in said District. Fourth. Whereupon the Recorder shall find it necessary to go to any location made by any person within this District before recording the same he shall receive from the applicant in addition to his fee for recording twenty-five cents per mile traveling fee in going to said location and five dollars per day or at that rate for part of the day. Fifth. The recorder shall give to all persons who have claims recorded in his office a certificate of the same when requested with a description of the location and all necessary facts to identify the same and shall receive one dollar for each certificate so made and signed by him.

Section Five. The provisions of an Act of Congress entitled an act to promote the development of the mining resources of the United States approved March 10th, 1873, is hereby adopted for the government of this Mining District in addition to the Rules and Regulations adopted at this meeting.

Section Six. The Rules and Regulations hereby adopted shall be in force from and after this date and shall not be altered or amended except at a meeting of the miners of the District called in conformity with the requirements of clause three, section four, and then only by a two-thirds vote of those present.

Section Seven. Every person who has an interest in any mine in this District or who is an actual miner within the District shall be a voter at a miners meeting provided such person is a citizen of the United States or a white citizen.

Section Eight. Any mine or mining locality lying contiguous to this district may be added to and become a part of this mining district by a request made in writing and signed by a majority of those owning or working said mine or in said mining locality, and a record of said application by the recorder of the District shall be all that is necessary to make said mine or mining locality a part of this District.

Section Nine. In case of the death, resignation or removal of the Recorder of the District any five persons who are voters may call a meeting to elect a new Recorder by posting three notices of a meeting for one week in three of the most public places within the District at which meeting a Recorder may be elected by a majority vote of those present.

On motion of Mr. Truman Mr. John E. Magee was elected to serve as Recorder for the Truman Mining District for the ensuing year. On motion of Mr. Hodge it was voted to hold an open meeting until 9 P.M. or such hour as may be necessary to complete the business of this meeting.

7 o'clock P.M. Mr. Hodge offered the following resolution which was adopted.

Resolved, that we recommend and urge upon all persons within this District that in all cases where differences may arise in reference to mining claims it is urged upon all interested to have the rights of parties decided by a jury of arbitration of twelve mining voters or less as may be agreed upon by the parties and that in no case will a countenance or assent to a resort to law or force in the settlement of mining rights until all reasonable actions have been made to procure a settlement of difficulties by arbitration as recommended above.

Mr. Hodge also offered the following resolution which was adopted.

Resolved, that the Recorder be and is hereby authorized to attach the names of those present at this meeting to the rules and regulations adopted and that he be instructed to keep a record of all persons now within or who may hereafter become voters within the District so far as he may be informed of the same and their locality.

On motion of Mr. Truman, the Minutes, Bye laws and

Regulations were ordered read. On motion of Mr. Hodge, the Bye laws, Regulations and Minutes were approved and the meeting adjourned sine die.

Signed H. C. Hodge, Henry Sims, Cornelius Ryan, J. C. Truman, A. C. Benedict, John Mansfield, Joseph King, Wm. G. Boyle, Chairman, John E. Magee, Secretary.

Filed and recorded at request of Wm. G. Boyle at 10 o'clock A.M., February 23 A.D. 1875.

W. Carpenter, Recorder

1875.03.17

Smith Mining District, Pima County, March 17, 1875 (Published in *The Citizen*, April 3, 1875, 4:2)

In pursuance of a written notice, a meeting of miners was held at Chalybeate Spring, Pima County, Arizona, on the 17th of March, 1875, for the purpose of organizing a mining district and enacting rules and regulations for the government of the same.

Said notice was duly posted at said place on the 12th day of the same month and year and signed by the following persons: D. B. Rea, Jerome Sawyer, W. Hanson, W. Gardner, Alva Smith, Jack Long, T. McArthur, P. J. Hand, W. Elliott, H. G. Morse, R. T. Stibell, J. W. Coughlan, R. M. Choate.

On motion of D. B. Rea,[6] Mr. Choate was elected chairman of the meeting, and W. Gardner appointed secretary.

On motion of Wm. Elliott, a committee of five persons were appointed to draft and report a code of by-laws, rules and regulations for the district. This committee was comprised of the following persons: T. McArthur, M. Stibell, W. Elliott, D. B. Rea and Jerome Sawyer, who reported the following rules and regulations, which were read and adopted by sections:

Sec. 1. This mining district shall be known as the Smith district, and shall be included within the following boundaries: commencing at old Camp Cameron and running to a point four miles east of Gardner's Ranch; thence to Davidson Spring; thence to Sierrita; thence to Camp Cameron.

Sec. 2. The annual meeting of the voters of this district shall be held on the 3d Monday in February of each year. At which meeting an election of a recorder shall be held, and such other business transacted as may be found necessary. Ten days previous notice shall be given of said meeting by the recorder posting three notices in three of the most public places within said district.

Sec. 3. There shall be elected at this meeting a recorder for the district who shall hold his office for one year from said date, whose duties shall be as follows:

FIRST. To file and record in his office all notices of location of lode and placer claims made within the boundaries of this mining district, and within every sixty days to have the same recorded in the county recorder's office; provided, the party so recording advance to him the of the county recorder for recording the same; and he shall have the custody of the books, the by-laws and proceedings of this mining district.

Sec. 4. Before recording any mining claim, the recorder shall inform himself by going in person to the claim or lode of every mining location presented for record; and shall examine the claim and see if there is an actual lead or deposit of mineral bearing rock or earth, in the boundaries claimed by the locator, and shall in no case make a record in favor of one party when another party to his knowledge claims the same location.

Sec. 5. The recorder shall call a meeting of the miners of this district upon the application of five miners of the district, and made in writing, and shall post, or cause to be posted, notice of the same for at least thirty days before the meeting, in three of the most public places in said district.

Sec. 6. The recorder shall receive two dollars for recording any notice of location.

Sec. 7. All placer or gold bearing gravel claims shall not exceed in extent one thousand feet in length and one hundred fifty in width to one person, and he shall not hold more than one such claim on one gulch by preemption, but shall hold as many as he pleases by purchase.

Sec. 8. In case of death, resignation or removal of the recorder of the district, any five legal voters of the district may call a meeting to elect a new recorder by posting three notices of a meeting to elect a recorder, in three public places in the district, for one week before such election. A recorder may be elected by a majority at such election.

Sec. 9. No one shall be allowed to vote in any district election except those who own or have an interest in a mining claim within the district.

Sec. 10. In all cases where disputes or differences shall arise as to the rights of mining claims, the rights of the parties shall first be submitted to an arbitration of seven qualified voters of the district, to be chosen by lot or a less number, if agreed to by the persons who shall sit and hear the evidence and arguments of the respective parties, or their attorneys, and give their decision thereon before the parties shall resort to law or force in the settlement of their rights. And in view of the vexations and ruinous litigation that so often arises over mining claims, every person interested in mining claims in this district, should urge upon each other this mode of settling their disputes.

Sec. 11. All claims shall be recorded within thirty days after location.

Sec. 12. The recorder shall give to all persons who have claims recorded in his office, a certificate of the same when requested, and shall receive one dollar for each certificate so made and signed by him.

Sec. 13. All claims shall be signalized by stakes firmly set in the ground, and mounds of earth thrown up around them for each corner, or by notices placed upon trees, or other objects sufficiently firm, and a written notice with full description of the claim posted at or near the original discovery shaft.

Sec. 14. The laws relative to the location, representation and holding quartz lode or tunnel claims, shall be the same as those of the United States, passed by Congress May 10, 1872, entitled "An Act to promote the development of the mining resources of the United States" and shall also be applicable to the representing and holding of placer or gravel claims.

Sec. 15. These by-laws shall take effect from and after the 25th day of March, 1875.

Nominations for district recorder being in order, R. H. Choate was nominated by William Gardner, and was thereupon elected unanimously to that office.

On motion of Mr. Alva Smith, a vote of thanks was tendered to the editor of THE ARIZONA CITIZEN for the kindly interest manifested by him towards this new mining camp.

On motion of Mr. Stibell, a vote of thanks was tendered to D. B. Rea, chairman of the committee on by-laws, for the faithful discharge of his duties.

On Motion of H. G. Morse, meeting adjourned sine die.

Wm. Gardner, Secretary

1875.08.10

Pioneer Mining District, Pinal County, August 10, 1875 (published in *The Arizona Citizen*, 1:2)

Pinal District

Minutes of miners' meeting held at the camp of the Silver King mine, August 10, 1875:

The miners met in regular convention in pursuance of notice given on this, the 10th day of August, 1875, at 2 o'clock p.m. Marshall Barnes, chairman, and Harry Jones, secretary.

The secretary read the minutes of the proceedings of the last meeting.

Messrs, C. O. Brown, M. Barnes, R. D. Dicky, Wm. G. Boyle and Richard Powers, having been appointed as a committee to draft resolutions, they handed in their report, which was acted upon by sections and adopted. The report as so adopted, was then recorded as follows, to-wit:

At a miners' meeting held at Silver King mine, the following local laws were adopted as the miners' laws of the district:

First. The name of the district shall be the Pioneer district, and its boundaries are as follows:

The northern boundary a line draw east and west and ten miles long, the center of which shall be ten miles due north of Silver King mine; the southern boundary, a line drawn east and west ten miles south of Silver King Mine; the east and west lines shall be drawn north and south to connect the north and south lines, constituting an area of ten by twenty miles, with the Silver King mine in the center.

Second. No variation, alteration, or modification of the liberal law of the United States shall be tolerated, its provisions being ample for all purposes.

Third. No local record of mining claims shall be kept and all notices shall be recorded In the county recorder's office of Pinal county.

It was further resolved that the chairman of this meeting cause a copy of these resolutions and by-laws to be recorded in the office of the county recorder of Pinal county.

A unanimous vote of thanks was tendered Mr. J. A. Devine in recognition of his having furnished this district with a list of all claims recorded therein.

There being no further business, the meeting adjourned sine die.

HARRY JONES, Secretary

1875.10.30

PECK MINING DISTRICT, YAVAPAI COUNTY, OCTOBER 30, 1875 (PUBLISHED NOVEMBER 12, 1875, IN *WEEKLY ARIZONA MINER*)

Pursuant to a call, made ten days previous by posting notices at the Peck camp, the Silver Prince camp, Lount's camp and Goodwin & McKinnon's camp, a miners' meeting was held at Ed Peck's house, near the Peck mine, on Saturday, October 23rd. No action was had and the meeting was adjourned to Saturday, Oct. 30.

Saturday, Oct. 30.—The adjourned meeting was called to order at the mess-house of the Peck Mining Company, at 2:30 o'clock p.m., and organized by the election of Geo. Hogle, chairman, and Henry A. Bigelow, secretary.

On motion, the chairman appointed George Opdyke and H. A. Bigelow a committee to draft resolutions, define boundaries of a new district, etc. On motion the chairman was added to the committee.

After the lapse of a little time and committee reported as follows:

Whereas: We, the miners in the neighborhood of what is known as the Peck mining camp, are unable to decide whether we are located in Turkey Creek district, Bradshaw Mountain district, Bradshaw, Pine Grove, or Agua Fria district , and as no organization has been in operation in either of said districts, or in any district adjoining, for a year past; and as all claims recently located have been taken up under the Congressional laws,

Therefore, we now proceed to make a new district, defined as follows: This shall be known as the Peck Mining District, and shall be bounded thus: Commencing at the mouth of Poland creek, at its junction with Turkey creek, thence running up Turkey creek to the mouth of Bear creek, to where Battle Flat creek empties into it; thence up the ridge, on the top of the same, between Bear creek and Battle Flat creek, to the top of Bradshaw mountain, at the head of said ridge, thence running due south to Poland creek, thence down Poland creek to the place of beginning.

That with the exception of regulations hereinafter specified, the Congressional laws of May 10th, 1872, and subsequent amendments thereto, shall be the laws of this district.

That the County Recorder of Yavapai county is hereby declared the recorder of this district.

That all claims already located in this district, and not recorded, shall be recorded within thirty days, and all claims hereafter located shall be recorded within thirty days from the date of location.

That all claims hereafter located in this district shall, in addition to the principal notice claiming the ground and stating the boundaries, have a brief notice in each of the corner monuments and centre end monuments stating which corner or end monument it is, and giving the name of the claim and the name or names of the owner or owners thereof, and we urge all owners of claims in this district to put such notices in their corner and end monument at as early a day as possible in order to clearly show to other miners and prospectors the limits of their respective claims.

All laws of any other mining district or districts which may have included within these boundaries a part or the whole of this district are, within the limits of this district, hereby repealed.

On motion, the report of the committee was accepted and the committee discharged.

On motion, the report of the committee was adopted as the action of the meeting, and the secretary was requested to furnish a copy of the proceedings to the Editor of the Arizona Miner for publication.

On motion of Mr. Opdyke the meeting adjourned sine die.

Geo. T. Hogle, Chairman
Henry A. Bigelow, Secretary

1875.11.25

Globe Mining District, Maricopa County, November 25, 1875 (Published December 17, 1875, in *Weekly Arizona Miner*, 1:3)

BOUNDARIES AND LAWS

Pursuant to notice, the miners in the neighborhood of the Globe Mine, on Pinal Creek, Maricopa county, Territory of Arizona, met together on the 25th of November 1875, at the camp of A. R. Hammond, for the purpose of forming a Mining District and making the necessary laws to govern the same. R. B. Metcalf [Medcalfe], was called to the chair and R. H. Choate, appointed secretary of the meeting.

The chairman stated the object of the meeting, and his remarks were followed by a motion, that the district about to be formed should be called the "Globe Mining District," which was carried.

A motion was also carried that the boundaries of the Globe District be as follows:

Commencing at a point where the supposed line of the San Carlos Indian Reservation crosses Salt River; thence running down along the course of said river to the mouth of Pinto Creek; thence in a southerly direction to the "Bloody Tanks" on the trail from the Globe mine to Pinal Post; thence along the summit of the timber range of the Pinal Mountains, to the Gila river; thence up the Gila River to the supposed line of the San Carlos Reservation; thence along said line to the place of beginning.

This District is intended to embrace that portion of the San Carlos Reservation that is about to be cut off and declared open for the occupation of miners and prospectors.

The motion carried that the Mining Act passed by Congress on May 10, 1872, and all subsequent amendments to that Act, be and are hereby adopted as the laws governing this district.

The following By-Laws and regulations were then adopted:

SECTION 1.—That each claim located in this district shall be recorded in the district records within thirty days after the date of Location, and any failure to comply with this rule shall be considered an abandonment, and the claim subject to relocation as though it had never been located.

SECTION 2.—There shall be upon each claim recorded in the district the amount of twenty-five dollars in labor expended during the first three months after the date of location.

SECTION 3.—Any failure to comply with the above rule, shall be deemed as abandonment of the claim.

SECTION 4.—All claims located in the district prior to this date, shall be recorded in the district records, within sixty days from the date of the passage of these laws.

SECTION 5.—Each notice of location presented for record, shall bear the signatures of one or more witnesses to the location.

SECTION 6.—There shall be notices posted in three conspicuous places in the district five days previous to the calling of any and all subsequent meetings of the miners of this district, and five mine owners in the district may call such meeting.

SECTION 7.—In all miners' meetings hereafter held in the district, fifteen mine owners in the district shall be considered a quorum to transact business, and a two-third vote of all mine owners present at any meeting shall be required to alter, abolish or make new laws for this district.

SECTION 8.—On the 25th of November of each year there shall be a meeting of the miners of this district, held for the purpose of electing a recorder, and for such other business as may be brought before it.

SECTION 9.—The recorder shall be elected to serve the term of one year.

SECTION 10.—It shall be the duty of the recorder to keep the books and papers of his office within this district.

SECTION 11.—The recorder shall be entitled to a fee of one dollar for recording each claim, and one dollar for each certificate furnished.

SECTION 12.—It shall be the duty of the recorder to appoint a deputy to act in his stead during his absence from the district.

A motion was carried that the secretary be instructed to forward a copy of these proceedings to the Arizona Citizen, the Prescott Miner and the Silver City Herald for publication.

Signed, R. B. METCALFE, Chairman
R. H. CHOATE, Secretary

C. M. SHANNON, I. WINTERS, M. MORRIS, JOHN HARVEY, N. L. GRIFFIN, F. TARBELL, A. R. HAMMOND, M. ENSENA, GEO. SCOTT, EDWIN POLK, C. E. BUCK, OMER WHITLOCK, G. C. NOLAND, L. E. JONES, F. J. MORRIS

R. H. CHOATE, Secretary

Meeting adjourned.

TERRITORY OF ARIZONA,

County of Gila, ss.

I, G. A. Swasey, District Recorder of Globe Mining District, County and Territory aforesaid, do hereby certify that the foregoing is a full, true and correct copy of the Laws, etc. of said Globe Mining District, as found recorded in the Records of said District, and that the same are now in full force and virtue.

Witness my hand this 25th day of February 1881.

G. A. SWASEY, District Recorder said Globe Mining District

1876.04.17

Verdi [Verde] Mining District, Yavapai County, April 17, 1876 (Recorded in District Records, pp. 92–94)

Miners Meeting

At a Miners Meeting held at S. V. Kell, on the old Indian reservation west of the Verdi River, Yavapai County Arizona territory on the 17th day of April 1876, for the purpose of beginning a mining District the following resolutions and By-Laws were proposed approved and adopted.

Resolutions

First. That the District shall be named the Verdi Mining District with boundaries as follows: Commencing at a point on the Verdi River known as Rose Cañon running from thence West for ten miles thence south ten miles thence East ten miles thence North ten miles to place of beginning.

Second. That a District mining Recorder be elected for Said District until the Annual election be held for the purpose of Electing a District Mining Recorder and that said Recorder shall perform the duties of said office as required by the Rules and Regulations mentioned in By Laws of the Verdi Mining District until said Annual election.

By Laws

First. That the Mining Laws of the Verdi Mining District shall be in entire conformity with the Act of Congress approved 10th May 1872 An Act entitled an Act to Promote the development of the Mining Resources of the United States.

Second. That an annual meeting will be held each year on the first Monday in April for the purpose of electing a Mining Recorder for said Verdi Mining District.

Third. That said Meeting mentioned in clause Second will be called by the Mining Recorder, by posting in three conspicuous places in the Mining District a notice that such a meeting shall be held for such purpose.

Fourth. That at said Annual Meeting the Recorder will be elected by the vote of the Miners assembled at said meeting and that any candidate receiving a majority of votes will be the legally elected District Recorder who will hold the office until the next annual election or until his successor is appointed and that no person shall be entitled to vote at said election unless a mining interest shall be owned in said District by the Voter.

Fifth. That it shall be the duty of the Mining Recorder to reside in the Mining District and that he shall keep Books in which he will enter all Mining Locations made in the District and the work (as required by said Act of Congress) done on Locations.

Sixth. That in case the Mining Recorder should be obligated to leave the district he will appoint during his absence a competent deputy to perform the duties of his office.

Seventh. That the recorder shall be entitled to a fee of two dollars and fifty cents for recording any location and should there be any more names in the location he shall not be entitled to any additional fee for any number of names. Recorder shall also be entitled a fee of fifty cents for recording the work done on a location and giving a certified Certificate of same to Locator who performed the work but in no case shall Recorder enter any record of work done on any location without having personal knowledge that the required work is done or having the testimony of two witness whom he knows, that said required work was done.

Eighth. That all work done on any ledge mines or location on the Verdi Mining District as required by Act of Congress shall be entered in the Books of the District Mining Recorder.

Ninth. That all Ledges or mining claims discovered in this District shall have notices placed on them on some conspicuous place on the mine defining from the point of location on said notice the number of feet claimed and in what direction from the notice said feet are claimed in Length, and said notice shall also state the number of feet in width along the vein or lode claimed by the locator and should Locator not mention the number of feet in width, only the minimum number of feet all by Act of Congress of May 10, 1872 shall locator be entitled to and that said notice shall be recorded in the Books of the District Mining Recorder within thirty days from the date of discovery of vein or lode.

Tenth. That the books of the District Mining Recorder shall be open for inspection at his office form 10 ocl A.M. until 4:00 P.M., on all week days, but in no case shall said books be inspected unless in the presence of the Recorder or his deputies.

Eleventh. That all claims or Mines located in the Verdi Mining District previous to the adoption of these by Laws shall not come under the rules and regulations of said By Laws until after the first Annual Election of April 1877.

After the foregoing Resolutions and By Laws were proposed and adopted the Miners assembled proceeded to elect a Recorder until the Annual Election of April 1877, When S. V. Kell was unanimously Elected.

J. D. Boyd, President
Edw O. Doughterty, Secretary

1876.11.17

Tyndall Mining District, Pima County, November 17, 1876 (Recorded November 17, 1876, in Book 1 of Misc. Records, pp. 394–97; January 15, 1877, in Book 1 of Misc. Records, pp. 607–8; and December 11, 1877, in Book 1 of Misc. Records, 511–12)

Organization of the Tyndall Mining District with Rules, Regulations and Byelaws of the same together with the minutes and proceedings of a miners meeting held at the Salero Mine in Pima County Arizona territory by virtue of a notice posted in said Tyndall Mining District in pursuance of law.

Notice

There will be a meeting of all the miners in this District for the purpose of ascertaining the necessity and feasibility of establishing or forming a mining District to include twenty (20) miles square embracing the Salero mine where this notice is posted. Said meeting will be held at the Salero Mine on the 17th day of November A.D. 1876 at 12 oclock M.

Signed Wm G. Boyle, John E. Magee, George Cooler, Thomas Davis. Posted on the ground November 16th, A.D. 1876.

Meeting

Tyndall Mining District Salero Mine Pima County Arizona November 17th 1876 at 12 oclock M. Notice of meeting was read and meeting called to order by Wm G. Boyle, present John E. Magee, Thomas Davis, George Cooler, Wm G. Boyle.

Minutes

On motion of Mr. Cooler, Mr. Boyle was elected chairman and called to the chair. On motion of Mr. Davis, Mr. Magee was elected secretary of the meeting. On motion of Mr. Magee the meeting resolved themselves into a committee of the whole for the purpose of making rules, Regulations and Bye Laws who unanimously Reported the following of which were accepted. "Rules Regulations and Bye Laws."

1st This Mining District shall be called the Tyndall Mining District.

2nd This District shall embrace the following territory to wit: Commencing at the highest Santa Rita Peak that is to say the highest Peak of the Santa Rita Mountains in the County of Pima, Arizona Territory and run thence west twelve (12) miles, thence south twenty (20) miles, thence east twelve (12) miles thence north twenty (20) miles to the place of beginning to wit: the highest peak of the said Santa Rita Mountains.

3rd The County Recorder of Pima County by virtue of his office shall be ex officio Recorder of this District.

4th In connection with the Rules Regulations and Bye Laws adopted at this meeting Chapter Six (6) of Title thirty two (32) of the revised statues of the United States are hereby adopted for the government of this Mining District.

5th In the location of mines in this District copies of the Notice of location must be placed on the Mine before the same is transmitted to the country Recorder for Record or the location will be void.

6th All location notices must be filed in the office of the Recorder within Sixty (60) days after the actual date of location.

7th The county Recorder shall be entitled to the sum of two ($2.00) dollars for each and every notice recorded by him

8th The Records of Pima County are hereby adopted as the Bona fide Record books of this District.

9th The annual meeting of the voters of this District shall take place and be holden on the last Monday in November in the year A.D. 1877 and on the same day in each subsequent year.

10th Ten days prior to the date of holding said meeting the Recorder shall cause to be posted in three of the most conspicuous places in said District a notice Stating the time when and the place where said meeting shall be holden and shall designate in such notice that the meeting will be holden for the purpose of transacting all and every kind of business which may be brought before it.

11th At each annual meeting the miners and voters of the District shall elect their chairman and Secretary who shall hold office for one year.

12th The chairman and Secretary this day elected shall hold office for one year or until the last Monday in November A.D. 1877.

13th The Secretary of each meeting shall keep full and complete record of the minutes and proceedings of their respective meetings and cause the same to be placed on record in the office of the county Recorder.

14th These Rule Regulations and Bye Laws shall not be altered or in any way changed except at a regular annual meeting of the miners of said District and legal vote of two thirds of all the voters present.

15th Any and every person who is a citizen of the United States or who has declared his intention in due form of law to become a citizen of the United States and who own a share or shares in an mine or mines in this district or who is at work for any person or persons who own a mine or mines on said mine or mines for at least twenty days prior to said meeting shall be considered a legal voter of said District and shall be entitled to vote at all the annual meetings of said District.

16th Five dollars per day shall be allowed for each and every days work of ten hours labor performed upon a mine for the purpose of holding title or perfecting the requisite amount of work for a Patent and no other cost for provisions tools transportation or any thing shall be considered as expended for such work on any mine.

17th These Rules Regulations & Bye Laws shall be filed and Recorded in the office of the County Recorder and shall be in force from and after this date to wit: the 17th day of November A.D. 1876.

18th The above and foregoing shall be signed by all miners present at this meeting and shall be deposited with the Recorder within 30 days. Wm

G. Boyle chairman, George Cooler, Thos Davis, John E. Magee. I hereby certify that the above Bye Laws rules and regulations were adopted and signed as above on this the 17th day of November 1876. John E. Magee, Secretary, filed and Recorded at request of John E. Magee, November 25th A.D. 1876 at 9 oclock A.M.

Sydney Carpenter, County Recorder

Meeting of December 28, 1876

Election of District Recorder, John E. Magee elected. Note that County Recorder, Sydney Carpenter was among the miners.

Meeting of November 26, 1877

Minutes of the Meeting, Salero Camp, Tyndall Mining District Pima County Arizona Territory, November 26th 1877, One Oclock P.M.

Meeting called to order by Secretary John E. Magee. Present Thos Davis, John Burns, Cornelius Ryan and John E. Magee.

On motion of Mr. Davis, Cornelius Ryan was elected Chairman for the ensuing year.

On Motion of Mr. Burns, John E. Magee was elected secretary for the ensuing year.

Mr. Magee presented the following Resolution:

Be it resolved by this meeting that Section 2nd of the Rules Regulations and Bye Laws of the Tyndall Mining District shall be changed to read as follows and same shall be in full force and effect from the adjournment of this meeting.

Sec 2nd This district shall embrace the following territory to wit: Commencing at the Easterly end of the Empress of India Mine in Tyndall Mine in the Tyndall Mining District and running thence Northerly to the Eastern end of the Georgia Mine in said district, thence Northerly along the western boundary of the Aztec Mining district to the North west Corner of said District, thence Northerly to the highest Peak of the Santa Rita Mountains, thence west Nine Miles. Thence easterly following a line ¼ of a mile east from the Eastern base of the San Calletano Mountains to a point due west from the Empress of India Mine and thence East to the point of commencement, the Eastern end of the Empress of India Mine.

Mr. Davis made a motion that the Resolution as offered by Mr. Magee be adopted, Seconded by Mr. Burns. Motion posed by the Chairman and adopted.

On Motion of Mr. Davis the meeting adjourned sine die.

I certify that the above is a correct copy of the Notice Elections, Resolutions and minutes of the first annual Meeting of the Miners of the Tyndall Mining District holden on November 26, 1877 at Salero House in said District Pima County, Arizona territory.

John E. Magee, Secretary

Filed and Recorded at request of John E. Magee, December 11th A.D. at 30 minutes past 11 oclock A.M.

Sydney Carpenter, County Recorder

1877.03.24

Pima Mining District, Pima County, March 24, 1877 (Recorded at Book 1 of Misc. Records at 425–28)

Organization of the Pima Mining district Pima County Arizona Territory the rules, regulations and by-laws of the same together with the minutes and proceedings of a miners meeting held at Camp known as Ruskville Pima Co. A.T. by virtue of a notice posted at said Ruskville in said District in pursuance of law.

"Notice"

There will be a meeting of the miners of this vicinity for the purpose of ascertaining the necessity and feasibility of Establishing or forming a Mining District to include ten (10) miles square embracing the Pima and adjoining mines—said meeting will be held at this place "Ruskville" on the 24th day of March A.D. 1877 at 12 O'Clock Noon.

Signed
Fred Drew
H. L. Lymons
A. Brichta

Dated on the ground is 14th day of March 1877.

"Meeting"

Pima Mining District Ruskville Pima County Arizona Territory March 24th 1877 12 o'clock noon.

The foregoing notice was read and the meeting called to order by Fred Drew. Present Fred Drew, H. S. Lymons.

"Minutes"

On motion of Mr. Brichta, Mr. Lymons was elected chairman and called to the chair. On motion of Mr Drew, Mr Brichta was elected Secretary of the minutes, on motion of Mr Drew the meeting resolved the members into a committee of the whole for the purpose of making rules and regulation and bylaws. The committee of the whole unanimously reported the following which were accepted and adopted unanimously as Rules, Regulations and Bye Laws.

1st This Mining District shall be called the Pima Mining District.

2nd This District shall embrace the following described and bounded territory. Commencing at a point two (2) miles north from Castle Rock running thence east two (2) miles, thence south ten (10) miles, thence west ten (10) miles thence north ten (10) miles thence East eight (8) miles to point of beginning.

3rd The County Recorder of Pima County by virtue of his office shall be ex officio Recorder of this District

4th Chapter six of Title Thirty two Revised Statutes of the United States is adopted as this article.

5th In the location of mines in this District copies of the notice of location must be placed on the mine before any legal record of the same can be made by the Recorder. Any location not so made shall be null and void.

6th All location notices must be filed in the office of the Recorder within sixty (60) days after the actual date of location.

7th The County Recorder shall be entitled to a fee of two (2) Dollars for each and every notice recorded by him.

8th The Records of Pima County are hereby adopted as the bona fide records of this District.

9th the annual meeting of the voters of this District shall take place and be holden on the first Monday in March in each year.

10th Ten days prior to the date of holding the annual meeting the Recorder shall cause to be posted in three of the most conspicuous places in said District a notice stating the time when and the place where said meting shall be holden and shall designate in said notice that the meeting will be holden for the purpose of transacting all and every kind of business which may be brought before it.

11th At each annual meeting the voters of the District shall elect their Chairman and secretary who shall hold office for one year or until their successors are appointed.

12th The Chairman and secretary of this meeting shall hold office until the next annual meeting or until their successors are Elected.

13th The Secretary of each meeting shall keep full and complete records of the minutes and proceedings of these respective meeting and cause the same to be placed of record in the office of the County Recorder.

14 These Rules regulations and Bye laws shall not be altered or in any way changed except at a regular annual meeting of the miners of said District and then only by a legal vote of two thirds 2/3 of all the voters present.

15th Any and all persons who are citizens of the United States or who have declared their intentions to become such and own shares of stock or interests in any mine in the district or who has worked in any mine in the District for the twenty days preceding such meeting shall be considered a legal voter and entitled to vote at a miners meeting.

16th Five (5) Dollars per day shall be allowed for each and every eight hours work performance upon a mine for the purpose of holding title or performing the necessary amount of work for a patent and no other expenses shall be considered as expenditures on any mine for the purpose of holding or perfecting title.

17th All mines hereafter located in this District shall be marked by end and corner monuments or stakes at least eighteen inches in height above ground with sufficient markings placed in or on them to designate which end or corner of the claim they designate. If stakes are used they must be sunk at least six (6) inches in the ground and have a blaze and figures upon one side.

18th All locations made and recorded previous to the adoption of these rules Regulations and Bye laws are hereby legalized so far as they may conflict with these laws.

19th These Rules Regulations and Bye Laws shall be filed and recorded in the office of the County Recorder of Pima County and shall be in full force and effect after this date 24th day of March 1877.

20th The above proceeding of and the proceedings of any subsequent meeting shall be signed by the chairman and secretary and transmitted by them to the Recorder as soon as possible.

Fred Drew, Miner
John B. Best
James M. Leggett
J. R. Baldwin
Gwen Rusk
Wm Manscoe

Pima Mining District
Pima County Arizona

We hereby certify to the correctness of the above minutes meeting and notice and that the above Rules Regulations & Bye Laws were adopted and signed on this 24th day of March 1877.

H. S. Lymons, President
A. Bricta, Secretary

Filed and recoded at request of Fred Drew March 26th A.D. 1877 at 11 o'clock A.M.

S. W. Carpenter, County Recorder

1877.04.29

Arivaca Mining District, Pima County, April 29, 1877 (Recorded in Book 1 of Misc. Records, p. 444, shown on Rand, McNally, USGS Report, and ABM Map)

By virtue of a Notice duly signed and posted April 21st, 1877, a meeting of miners was held at the Van Nosta and Camp on the 29th of April, 1877. The object of the meeting being the organization and formation of the Arivaca Mining District. John McCafferty acted as Chairman Capt. E. P. Voisard as Secretary.

Following are the laws passed to govern the District.

1. The Mining District shall be called the Arivaca Mining District.

2. This District shall embrace the following described and bounded Territory: Commencing at a monumented stake surrounded by stones near the buildings of the Arivaca Ranch and running thence 10' West of South to the Sonora line, again commencing at the monumented stake and running thence East 10 miles which is the East Corner, thence North 15 miles which is the North East Corner, thence North West to the intersection of the Main road from Tucson to El Plomo, which is the North West Corner, and thence along said road to the Sonora line.

3. A District Recorder shall be elected for this district at each annual meeting.

4. Chapter Six of Title thirty two Revised Statutes of the United States is adopted as this article.

5. In the location of mines in this district, copies of the notice of location must be filed in the district before any legal record of the same can be made by the recorder. Any location not so made shall be null and void.

6. All notices of location of mines must be filed in the office of the District Recorder within thirty (30) days after the actual date of location.

7. The District Recorder shall be entitled to a fee of Two Dollars for each and every notice recorded by him.

8. The records of the Arivaca District shall be adopted as the bona fide Records of the district.

9. The annual meetings of the voters of this district shall take place and be holden on the last Monday of April in each year.

10. Ten days prior to the date of holding the annual meeting the recorder shall cause to be posted in three of the most conspicuous places in said district a notice stating the time when and the place where said meeting shall be holden and shall designate in such notice that the meeting will be holden for the purpose of transacting all and every kind of business which may be brought before it.

11. At each annual meeting the voters of the District shall elect their chairman and secretary, who shall as well as the District Recorder hold their office for one year or until their successors are appointed.

12. The Chairman, Secretary and District Recorder elected at this meeting shall hold their respective offices until the next annual meeting or until their successors are elected.

13. The Secretary of each meeting shall keep full and complete records of the minutes and cause the same to be placed on record in the office of the County Recorder.

14. The District Recorder must reside in the district in case of absence he may appoint a deputy, but in case of his departure from the district a special meeting of the voters of the district shall be called and his successor elected.

15. Any and all persons who are citizens of the United States or who have declared their intention to become such and owns shares of stock or interest in any mine in the district or who has worked in any mine in the district for the twenty days preceding such meeting shall be considered a legal voter and entitled to vote at a miners meeting.

16. Five Dollars per day shall be allowed for each and every eight hours work performed upon a mine for the purpose of holding title or performing the necessary amount of work for a Patent, and no other expenses shall be considered as expended on any mine for the purpose of holding or perfecting title.

17. All mines hereafter located in this district shall be marked by end and corner monuments or stakes at least eighteen inches in height above

ground with sufficient marks placed in or on them to designate; if stakes are used they must be sunk at least twelve inches (12 in.) in the ground and have a blaze and figures upon one side.

18. All locations heretofore made and recorded previous to the adoption of these rules, regulations and bylaws are hereby legalized as far as they may not conflict with the same.

19. These rules, regulations and bylaws shall be filed and recorded in the office of the County Recorder of Pima County, and shall be in full force and effect from and after this 28th day of April, 1877.

20. The above proceedings and the proceedings of any subsequent meetings shall be signed by the Chairman and Secretary and be submitted by them to the Recorder as soon as possible.

21. These rules, regulations and bylaws shall not be altered or in any way changed except at a regular annual meeting of the miners of this district, and then only by a legal vote of two-thirds of all the voters present.

22. No more than one location shall be taken up by any one person on any one lode.

John McCafferty, Miner
E. P. Voisard, Miner
Ed Hudson, Miner
Wm. G. Poindexter, Miner
Jay Holden, Miner
Attest E. P. Voisard, Secretary

Filed and Recorded at request of E. P. Voisard—May 4, 1877, at 9 o'clock a.m.
S. W. Carpenter, County Recorder, By J. A. Apperson, Deputy

1877.05.06

Aztec Mining District, Pima County, May 6, 1877 (Recorded in Book 1 Misc. of Records, pp. 449 and 685, shown on Rand, McNally)

By virtue of a notice duly signed and posted on May 6th, 1877, a meeting of miners was held at place of posting notice in the 9th Instant at which place and time the Aztec Mining District was formed.

Mr. Thomas Davis acted as Chairman and Col. B. J. Hinton as Secretary. Following are the laws passed to govern the District.

1. This Mining District shall be called the Aztec Mining District.

2. This District shall embrace the following described and bounded Territory:

Commencing at the Easterly end of the Empress of India Mine and running west of north along the Eastern boundary of the Tyndall Mining District to the Eastern end of the Georgia Mine in the Tyndall Mining District Santa Rita Mountains, thence due North two miles, thence due East three miles, thence South six miles, thence Westerly to the point of starting at the Eastern end of the Empress of India Mine in the Tyndall Mining District Santa Rita Mountains.

3. The County Recorder of Pima County by virtue of his office shall be an official Recorder of this District.

4. Chapter Six (6) of Title Thirty-two (32) Revised Statutes of the United States is adopted as this article.

5. In the location of mines in this District copies of the notice of location must be placed on the mines before any legal record of the same can be made by the Recorder. Any location not so made shall be null and void.

6. All location notices must be filed in the office of the Recorder within thirty (30) days after the actual date of location.

7. The County Recorder shall be entitled to a fee of Two Dollars ($2.00) for each and every notice recorded by him.

8. The records of Pima County are hereby adopted as the bona fide records of the District.

9. The annual meeting of the voters of this District shall take place and be held on the first Monday in May in each year.

10. Ten days prior to the date of holding the annual meeting the Recorder shall place or cause to be placed or posted in three (3) of the most conspicuous places in said District, a notice stating the time when and the place where such meeting shall be held, and shall designate in such notice that the meeting shall be held for the purpose of transacting all and every kind of business which may be properly brought before it.

11. At each annual meeting the voters of the District shall elect their Chairman and Secretary who shall hold office for one year or until their successors are appointed.

12. The Chairman and Secretary of this meeting shall hold office from the ninth day of May for one year or until their successors are elected.

13. The Secretary of each meeting shall record full and complete records of the minutes and proceedings of these constructive meetings and cause the same to be placed on record in the office of the County Recorder.

14. These Rules, Regulations and Bye laws shall not be altered or in any way changed unless at a regular annual meeting of the miners of said District and then only by a legal vote of two-thirds of all the voters present and voting.

15. Any and all persons who are Citizens of the United States of America or who have declared their intentions to become such and own shares of stock or interest in any mine in the District or who have worked in any mine in the District for the twenty (20) days preceding such meeting shall be considered a legal voter and entitled to vote at a miners meeting.

16. Five dollars ($5.00) per day shall be allowed for each and every eight (8) hours work performed upon a mine for the purpose of holding title or performing the necessary amount of work for a Patent and no other expenses shall be considered as expended for the purpose of holding or perfecting title.

17. All mines hereafter located in this District shall be marked by end and corner monuments of stakes at least eighteen (18) inches in height above ground with sufficient marks placed of the claim must be sunk in or upon them to show which end or corner they designate. If stakes are used they must be at least six (6) inches in the ground and have a blaze and figures on one side.

18. All locations made and recorded previous to the adoption of these Rules, Regulations and Bye laws are hereby legalized as far as this may not conflict with the same.

19. These Rules, Regulations and Bye laws shall be filed and recorded in the office of the County Recorder of Pima County and shall be in full force and effect from and after this ninth day of May 1877.

20. The above proceedings and the proceedings of any subsequent meeting shall be signed by the Chairman and Secretary and submitted by them to the County Recorder without delay.

(Signed) Thomas Davis
John Mansfield
Wm. G. Boyle
Arthur S. Lovelock
John E. Magee

I certify that the foregoing is a correct statement of the proceedings had as of the laws adopted for the Aztec Mining District this ninth day of May, 1877.
Richard J. Hinton, Secretary

Filed and Recorded at request of John E. Magee May 12, 1877, at 4 o'clock P.M.

S. W. Carpenter, County Recorder
By J. A. Apperson, Deputy

1878.04.06

Tomb Stone Mining District, Pima County, April 6, 1878 (Recorded in Book 1 of Misc. Records, pp. 526–27)[7]

Notice. We the undersigned having discovered some mineral bearing quartz ledges which we are desirous under the provisions of law to possess and work with a view to proffit and ownership. Do in order to establish the location of the claim and comply to the Customs of the County hereby create a new mining District to be called the Tomb Stone Mining District embracing within its boundaries all of that portion of Pima County enclosed by the following lines. Commencing on the San Pedro River at a point opposite the old mines and about eighteen miles above the upper crossing of the San Pedro River. Thence East to the western base of the Dragoon Mountains. Thence Southerly along the base of said Mountains continuing the line to the Mule Mountains. Thence West to the San Pedro River. Thence northerly down said River to the place of beginning including all that range of hills known as the Tomb Stone hills together with there [?]. The laws of Congress of the United States and of the Territory of Arizona shall regulate the locating and holding of Mining Claims in this District With the exception that parties finding a Mineral cropping shall be allowed thirty (30) days in which to establish and fix the boundaries of their claims. Estbl April 6th 1875 in Tomb Stone District Pima County Arizona (Signed)

A. E. Scheiffelin
Richard Gird
E. L. Scheiffelin
Oliver Boyer
Theo E. Walker

Filed and recorded at request of Richard Gird. April 9th, A.D. 1878 at 3 o'clock P.M.

S. W. Carpenter
County Recorder

1878.04.16

Helvetia Mining District, Pima County, April 16, 1878 (Recorded in Book 1 of Misc. Records, pp. 530–31)

Helvetia Mining District organization of Santa Rita Mountains, Mescal Ranch, Pima County Arizona. April 16th 1878. We the undersigned being desirous of forming a new Mining District do hereby declare:

1st That this District shall be known as the Helvetia Mining District.

2nd This District shall embrace all the following lands having an area of ten miles square taking for a center point of starting the Santa Rita Chief Mine.

3rd The Act of Congress approved May 10th 1872 shall constitute the By Laws & Regulations of the District.

4th The County Recorder shall be ex officio Recorder of this District and shall be entitled to the sum of 1 ($ one) dollars for recording each and every mining notice for new mines located in the District.

5th, the Records of Pima County shall be the bona fide Records of this District.

6th, The Miners of this District may unilaterally change these By Laws & Regulations at any regular meeting, Provided that twenty days notice of such meeting shall be given by posting notices in 3 conspicuous places in said District and having the same published in any newspaper published in this Pima County, and then only by a two thirds vote of all Miners Present.

Done at the Helvetia Mining District this 16th of April AD 1878.

B. Kifle [?]
Fred Hilfe [?]
Albert Tibelel [?]
William Mallen [?]
Thomas Billlew [?]
Thomas Redlick [?]

Filed and recorded at the request of B. Melli April 23rd, 1878 at 3 o'clock P.M. S. W. Carpenter, County Recorder by HA. Lemeoo, Deputy.

1878.04.29

Harshaw Mining District, Pima County, April 29, 1878 (Recorded May 2, 1878, in Book 1 of Misc. Records, p. 545)

We the undersigned miners being desirous of forming a new Mining

District with Bye Laws Rules and Regulation for the government of the same do hereby declare:

1st This District shall be known hereafter as the Harshaw Mining District.

2nd This District shall embrace all of the following Territory, to wit: An area of twelve miles square taking the center of the French Mill Site as a center point.

3rd That Revised Statutes of the United States Chapter six title thirty two and the laws of the Territory of Arizona are hereby adopted as the Mining Laws of this District.

4th The Country Recorder shall be ex officio Recorder of this District and the Country Records are hereby declared to be the bona fide Records of this District.

5th The Recorder shall be allowed the sum of two (2.00) Dollars for Recording the Location Notices of any mine in this District.

6th These Rules Regulations and Bye Laws shall not be changed except at a regular miners meeting by posting notice of such meeting in three of the most conspicuous places in the District and by Publication of such notice in a newspaper published in this Pima County and then only by a two third vote of the miners present.

Signed by all the Miners present this 12th day of April A.D. 1878. H. H. Holt, J. W. Davis, J. C. Otis, D. T. Harshaw, S. W. Carpenter, C. N. Labance, Daniel P. Tomey, Murdock McKay, Harry Clay, C. T. Dunavier, James Tea, Michael Fagan.

Filed and recorded at request of J.W. Davis April 30th A.D. 1878.

S. W. Carpenter, County Recorder

1878.04.30

Tucson Mining District, Pima County, April 30, 1878 (Recorded May 3, 1878, in Book 1 of Misc. Records, p. 548)

Camp Sacramento San Calletano Mountains, Pima County A.T., April 30th 1878.

At a meeting of Miners held at the above named Camp April 30th 1878, at which Chas. Bell, A. C. Benedict, J. Magee, E. A. Chambelain, W. L. Campbell, W. L. Wakfield, Frank Sullivan, J. W. Everts, Joe Plummer, D. P. Lowell and J. E. Tallmadge were present. Chas Bell was chosen President and W. L. Campbell Recorder.

It was resolved to form a mining District to be named the Tucson Mining District of Arizona to comprise the Territory bounded as follows:

On the east, by a line commencing at a point one mile and a quarter from the North eastern base of the San Calletano Mountains thence running southerly along the Eastern base of the San Calletano Mountains to a point on the Sonoita Creek at or near Sanfords. Thence along the top of the Patagonia Range to San Antonia Pass on the south by the boundary line between Arizona and Sonora running westerly from San Antonio Pass to a point on the summit of the Pajero Mountains.

On the west by a line running Easterly from said peak to the place of beginning. Adjourned to meet at Camp Sacramento June 1st 1878, for the purpose of adopting a constitution and Bye Laws for governing the District and transacting such other business as may come properly before the Meeting.

All miners within the Limits of the District are Notified to be present and assist in the completion of the organization.

Chas Bell, President
W. L. Campbell, Recorder

Filed and Recorded at request of Chas Bell May 3rd A.D. 1878 at 10 oclock A.M.

S. W. Carpenter, County Recorder

1878.07.22

Old Hat Mining District, July 22, 1878, Pima and Pinal Counties (Recorded on July 29, 1878, in Book 1 of Misc. Records of Pima County, pp. 561–62)

Pima County Arizona Territory July 22nd 1878. At a Miners Meeting held at the American Flag Mill Site Pima County AT (on the day above written) for the purpose of forming a mining District of which Meeting Louis DuPuy was appointed chairman and W. G. Guild Secretary the following Resolutions were adopted.

1st That the District be bounded as follows: to commence at Old Camp Grant on the San Pedro River that point to form the North East Corner of the District, thence following the Course of the River upst[ream] in a southerly direction Twenty-four miles thence in a westerly direction to the summit of the St Catheryn Mountain following the summit of the Mountain until the line strikes the Old Road from Tucson to Camp Grant thence following the Old Road to Camp Grant the place of beginning this mining District shall be known as the Old Hat Mining District.

2nd And it is further Resolved that all of the Records of said District be rec'd in the County Recorders Office at Tucson Pima County Arizona Territory.

3rd and further Resolved that the District be governed entirely by the United States Mining and that no local rules be adopted at Present.

Signed
Louis DuPuy Chirman
W. G. Guild
Wm Haueck
W. C. Deen
Issac Kaurin
John Foutes
W. H. Timmens

Filed and Recorded at request of Louis DuPuy July 29th A.D.1878 at 9 A.M.
S. W. Carpenter
County Recorder

[*Note*: It would have been important to designate a place of recording since the district overlapped two counties.]

1878.09.07

California Mining District, Pima County, September 7, 1878 (From recorder's book on file with BLM)

Proceedings of a Miners Meeting held at Leroy's Camp Silver Creek, Pima County, Arizona on the 7th day of September 1878.

By virtue of a notice duly signed and posted at three prominent places in a proposed mining District, dated August 24th 1878 a Meeting of the miners was held at the above named camp on the date first above written. Mr C. A. Milner was chosen Chairman and David P Carr Secretary of the Meeting. A Mining District was then proposed and organized to be called and known as the

California Mining District

Which mining district embraces and includes the following described and bounded territory to wit: Beginning at a point on the South bank of

Cave Creek due north of the most easterly point of the base of the hill east of Leroys Mill Site at the Sulpher Spring, and running thence due South four miles, thence due West to the summit of the Chiracahua Mountain thence Northerly along the summit of said mountain to the south line of the Chiricahua Mining District, thence Easterly along the said south line to a point distant one mile east of where said line crosses the wagon road passing up Turkey Creek, thence Southerly a direct line to the place of beginning.

The following were agreed upon and adopted as the laws and regulations to govern the said mining district.

I Chapter six of title thirty-two revised Statutes of the United States, the general Mining Laws of May 10th 1872 was adopted as the organic law of this district.

II All location notices on claim located prior to September 7th 1878 must be filed for record in the office of the District Recorder within thirty days from that date, and all location notices on claims located after that date shall be filed in the Recorder's office for record within thirty (days) after the actual date of location.

III All location notices on lode-claims before being recorded shall bear the signature of two witnesses who shall certify to the posting of the notice upon the claim the discovery of a valuable mineral in a vein lode or deposit and placing of monuments or stakes at the corners of the claim.

IV Not less than fifty dollars shall be expended in work upon each and every lode claim in this district within six months after the date of location, and same amount within the first six months of each year after such date of location and the remainder of the one hundred dollars required by the United States law shall be expended in work upon each and every lode claim in this district before the expiration of the last six months of each year after such date of location, Provided that the whole amount required by the United States law upon each and every lode-claim within each year after the date of location may be expended within the first six months of any year and any amount in excess of fifty dollars expended within the first six months of any year after the date of location upon any mining claim shall be deducted from the fifty dollars required to be expended thereon in the last six months of the same year, any and all lode claims in this district upon which not less than fifty dollars shall not have been expended in work within six months after the date of location or within the first six months of any year after such date, shall be subject to relocation.

V When any owner or owners of a lode claim shall have performed the work required by the forgoing section he or they shall from the expiration of the time specified in said section report the same to the District Recorder who shall require such proof of the performance of the work as he may deem sufficient and he shall visit and inspect the work upon the claim and if satisfied that the requirements of the law have been compiled with he shall thereupon issue a certificate of the performance of the work consist and what amount, if any in excess of fifty dollars had been expended thereon if within the first six months of any year after the date of location, which certificate shall first be recorded in the records of the district and then posted upon he claim at the point at which the principal part of the work shall have been done.

VI Five dollars per day for each and every eight hours work shall be allowed on account of the expenditure required upon mining claims in this district.

VII The district recorder shall be entitled to a fee of one dollar for recording each and every notice recorded and shall also be entitled to a fee of one dollar for issuing and recording each and every certificate of the performance of the work upon lode claims and a mileage at the rate of ten cents per mile to and from his office to each claim examined prior to issuing such certificate, the Recorder shall also be entitled to a fee of one dollar for recording each and every paper other than the above mentioned containing not more than one hundred words and an additional fee of fifty cents for each one hundred works and part thereof exceeding one hundred.

VIII The District Recorder this day elected shall serve for one year from and after this date and until his successor shall be elected.

IX The Annual Meeting of the miners of this district shall be held on the first Tuesday after the first Monday in September of each year at which meeting a recorder shall be elected to service for the ensuing year, the recorder shall be elected by ballot and the person receiving the highest number of votes shall be declared elected

X The District recorder shall give notice by posting notices duly signed by him, in three of the most conspicuous places in the district of the time and place of holding such annual meeting at least ten days prior thereto.

XI Any and all persons who are citizens of the United States or declared their intention to become such and who own any interest in any mining claim in this district, or who have worked in any mine therein, for the twenty days preceding such meeting shall be entitled to vote at such annual meeting.

XII These laws cannot be changed in any manner, excepting at the annual meeting, and then only by a vote of two thirds in favor thereof of the miners entitled to vote then present.

XIII the foregoing laws shall be in full force and effect from and after this 7th day of September 1878.

David P. Carr, having received a majority of all the votes cast by ballot for the office of District Recorder was declared elected to that office for the ensuing year.

It was resolved that the District Recorder send copies of the meeting to the Arizona Star at Tucson A.T. and the Grant County Herald at Silver City N.M. with request for publication.

The meeting then adjourned sine die.

C. A. Milner, Chairman
David P. Carr, Secretary

Meeting of September 7, 1880

At the last annual meeting held 7th day of September 1880 Section (4) four was repealed

And Section (5) five was amended to that said section now reads thus: That the Recorder shall at the request of any mine owner or his Heirs or Assigns fulfill all of the requirements of said fifth section.

1878.11.23

Catarina (Catalina) Mining District, Pima and Pinal Counties, pre–November 23, 1878 (Referred to in the *Arizona Sentinel*, November 23, 1878, 2:2–3)

The newspaper article referred to the creation of the district as 40 miles south of Florence, promoted by M. A. Baldwin, R. F. Straine, Colonel Bixby,

Colonel Scott and Prof. J. B. Caldwell in association with the Old Hat and American Flag Districts.

1878.12.26

Cochise Mining District, Pima County (Recorded in Book 1 of Misc. Records, p. 604)

Rocky Spring

December 26, 1878

At a meeting of miners held at Rocky Spring, Dragoon Pass, Pima County, Territory of Arizona, at which the following were present: G. W. Seay, P. Rusk, J. Taylor, J. Wool, M. Volmar, J. D. Fink and A. J. Mitchell. The following motions were carried and thus formed.

1st. The District shall be known as the Cochise Mining District and shall be bounded as follows:

Starting on east side of Small Pass about 5 miles East of Rocky Spring or Fink Mill Site and running thence Northerly along east of Dragoon Mountain about 12 miles to Pass Alamos and Silver City Road. Thence westerly along same road 6 miles. Thence South about 12 miles to Butterfields Road. Thence Easterly about 6 miles to point of commencement.

The following laws are to be observed within the boundaries of the above mentioned District:

1st The discoverer of a ledge or lode may erect one monument and shall be allocated 30 days to define his boundaries.

2nd The notice must be recorded within 60 days from date of location.

3rd The assessment shall be shaft 10 feet deep 6 x 4. The value of which shall be considered $100.00.

4th Any miner may locate more than one claim on same lead providing he does $100.00 worth of work on each claim.

5th The wages of a miner shall be $6.00 per day and fund all tools, powder, rations and necessary to complete his work.

George W. Seay
J. Taylor
M. Vollmer
A. J. Mitchell
P. Rusk
J. Wool
J. D. Fink

Dated December 26, 1878

Filed and recorded at request of A. J. Mitchell January 7, 1879, at 10:45 A.M.

S. W. Carpenter, County Recorder

By W. A. McDermott, Deputy County Recorder

1879.01.20

Silver Mining District, Yuma County, January 20, 1879 (Recorded in Book 1 of Misc. Records, pp. 303–4)

Miners meeting called by the miners of Silver District, Yuma County, Arizona, January 20th, 1879.

Warren Hammond, Chairman; C. F. Birnham, Secretary.

Resolutions adopted as follows:

Resolved: That we recognize the name of our District as Silver District.

Resolved: That the boundary lines of said District be defined as follows: commencing at the mouth of Charcoal Wash, on the Colorado River; thence into the Colorado River to Light House rock; thence in an Easterly direction to Ehrenberg Trail; thence Southerly to the Charcoal Wash, to the place of beginning.

Resolved: That the District adopt the United States Laws in regard to locating mining claims, and governing the same.

Resolved: To recommend all disputes arising from locating mining claims, and also all other disputes, to an arbitration of three men in the said District.

Resolved: To recommend a Recorder for the said District. Walter Millar was then nominated and duly elected.

Resolved: That the time given for recording claims shall be thirty days from date of location.

The meeting voted unanimously that no Chinamen be allowed to work in the District.

Resolved: That the landing at the Colorado River shall be known as "Norton's Landing."

Resolved: That these laws shall not be changed without a notice of ten days being posted in three conspicuous places in the said District, signed by five claim-owners of said District.

The meeting adjourned without day.

C. F. Birnham, Secretary

Recorded at request of Walter Millar, January 20th A.D. 1879 at 2 P.M.
George Lyng Recorder
By C. D. Lyng Deputy Recorder

1879.02.11

Plomosa Mining District, Yuma County, February 11, 1879 (Published on February 22, 1879, 2:1, and April 5, 1879, 3:4, in the *Arizona Sentinel*)

Plomosa District.

At a meeting of the miners of Plomosa District, Feb. 11th 1879, Geo. A. Ellsworth was appointed President pro tem, and C. L. Bancroft, Secretary, and the following resolutions adopted:

Resolved, that we recognize the name of our District as Plomosa District.

Resolved, that the location of the boundaries lines of said District shall commence at Centennial Well, on the Ehrenberg and Prescott wagon road, and thence run in a westerly direction to Round Mountain about five miles west of Granite Wash, thence in a south easterly direction to the Eagle Tail Mountain; thence northwesterly to the place of beginning.

Resolved, that the District adopt the United States Mining Laws in regard to locating claims and governing the same.

Resolved, that a Recorder be elected for the District for one year. On motion Mr. George A. Ellsworth was unanimously elected Recorder.

Resolved, that the time for recording claims shall be thirty days from the date of location.

Resolved, that the Recorder shall be allowed a fee of two dollars for each claim recorded.

Resolved, that these laws shall not be changed except at a regularly called miners meeting, notice of which shall be posted for ten days in three conspicuous places and signed by at least five claim owners in the said District.

Resolved that a copy of these resolutions be forward to the Arizona Sentinel and the Arizona Miner for publication.

Resolved, that the meeting adjourn sine die.

G. L. Bancroft, Secretary

Meeting on March 15, 1879

A copy of the proceedings at a miners meeting held in Plomosa Mining District March 15th, has been sent use for publication, we regret that its length prevents our giving it in full. One resolution should be followed by all Districts, viz that the claims be distinctly marked by stone monuments at least three feet high erected at each end of the claim and at each corner.

The value of assessment work is fixed at $12 per foot in blasting ground and $6 per foot in picking ground.

Copy of Notice of location to be recorded within twenty days with the Recorder of the District. J. W. Johnson was elected Recorder for one year: hereafter a Recorder is to be elected annually.

Peter Smith acted as Chairman of the meeting and J.W. Johnson as Secretary.

1879.04.28

Greenlee Gold Mountain Mining District, Apache County, April 28, 1879 (Recorded July 15, 1879, in Book A of Mining Records, p. 23)

April 28th 1879 Apache County, A.T.

At a miners meeting held at the Gold Camp on the Frisco River about six miles above Clifton for the purpose of forming a new Mining District; the following laws and regulations were made,

Resolved that a committee of four be appointed to draft laws and define the boundaries of said district.

Boundaries as follows: commencing at H. Lesinsky's dam about two miles above Clifton and running on a line Northwest to Eagle River, thence up Eagle River ten miles, thence to the mouth of the Bonita River, thence to Sunset Peak, thence to the place of beginning. The committee of four consisting of Mr Chapman, Mr. Bentz, Mike Fracy and J. H. Dorsey. Mr. I. D. Pinkard chairman, Mr. John Brick Sec.

Resolved that all mines shall be recorded within thirty days after location, or be subject to relocation.

Resolved, that the recorder shall send a transcript of his records to the County Recorder every sixty days.

Resolved that the recorder shall receive $1.00 for recording each claim.

Resolved that the recorder shall act as Sec. at all meetings.

Resolved that a copy of these proceedings be sent to Silver City, also to the Silver Belt at Globe City.

Resolved that $6.00 per day be allowed for assessment work on mines;

The meeting adjourned.

Office of Recorder Apache Co. A.T.

I hereby certify that the foregoing copy of By laws of Greenlee Gold Mountain Mining District were filed for record in my office at St. Johns the 15th day of July at 10 Oclock A.M. and that the forgoing By-laws recorded on Page 23 Book A Mining Records of Apache Co.

Richard J. Bailey, County Recorder
Per Chas Kimmear, Dept. Recorder

1879.04.30

Red Rock Mining District, Pima County, April 30, 1879 (Recorded May 7, 1879, in Book 1 of Misc. Records, pp. 658–59)

At a meeting held in Red Rock Cañon about 5 Miles Northeast from the North end of the point of Mountains, Pima County, Arizona on April 30th 1879 the following Miners are present: J. M. Williams, O. J. Reed, M. C. James, Thomas J. Ward, Thos Cusack, Chas Van Valkenberg, John L. Lount and David B. Rae.

The Meeting was called to order and J. M. Williams was elected President and David R. Rae secretary by a unanimous vote.

The following resolutions were unanimously accepted.

Resolved 1st That this district of called red Rock Mining District and that its boundaries be described as follows: Commencing at the center Monument of the Kitty Mines which is situated on [indecipherable] about eight (8) Miles Southeast from the Souaita [?], Thence running east five miles; Thence running North ten miles, Thence running west ten Miles—thence running South ten Miles, Thence running East five Miles forming a district ten miles square.

Resolved 2nd That claims located in this district may equal fifteen hundred linear feet in length along the course of the lead lode or Vein and three hundred feet on each side of the Middle of the said lead, lode or Vein and must be distinctly marked by not less than six (6) conspicuous stone monument and that each such Monument be designated by a slip of paper stating also upon which Monument the Notice May be found—

Resolution 3rd That all locations made in this district must be recorded in the County Recorder's office in Tucson within sixty days after the date of location.

Resolved 4th That the Meeting adjoined.

J. M. Williams chairman
O. J. Reed
Alex N. Childes
M. C. James
L. B. Dunne
Thos J. Ward
Mather Crocks
Chas Van Valkenberg
Benett Roth
John L. Lount
Henry Grinell
Thos Culsack
David B. Rae—Secretary

Filed and recorded at the request of

May 7th 1879 at 9.30 A.M.

S. W. Carpenter Recorder
By M. S. Makeky, Deputy

1879.08.03

Hartford Mining District, Pima County, August 3, 1879 (recorded in Book 1 of Misc. Records, p. 713)

Mining Regulations of Hartford District. By virtue of notices being duly posted on 3rd day of August, 1879, a meeting of miners was held in Ramsey Canyon on the 8th inst. at which place and time the Hartford Mining District was founded. Mr. S. P. Overton acted as Chairman and C. W. Russell acted as Secretary. Following are the laws passed to govern this District.

1st. The Mining District shall be called the Hartford Mining District.

2nd. The District shall be six miles square, the center of the District shall be at the North end of the discovery claim known as the Hartford Mine.

3rd. The Mining laws of the United States passed in 1872 with their amendments shall be adopted as a whole.

4th. That all discoveries of mines shall be properly monumented within twenty days after date of discovery and that sixty days shall be allowed after date of discovery for recording said claim.

5th. That all claims on each monument shall state the name of claim, date of location and what monument such notice is posted on.

6th. All locations made prior to this date the locators shall have fifteen days to comply with the foregoing Regulations.

7th. Five dollars per day shall be allowed for each and every 8 hours work per founder per a mine for the purpose of holding title. No other expenses shall be considered.

8th. The Recorder of Pima County are hereby adopted as the bona fide Recorder of this District.

9th. For the purpose of calling a miners meeting thirty days notice shall be given by posting three notices in the most conspicuous places in said District signed by not less than ten mine owners of this District.

10th. The proceeding of this meeting shall be filed and recorded in the office of the County Recorder of Pima County and put listed in the Arizona Citizen. This District is located in Ramsey Canyon in the Huachuca Mountains about 8 miles nearly south from Fort Huachuca, Pima County, Arizona.

S. P. Overton, Chairman
C. W. Russell, Secretary

Filed and recorded at request of C. W. Russell, Sept. 12, A.D. 1879 at 1:30 A.M.
S. W. Carpenter, County Recorder by Leo A. Reed, Deputy

1880.02.21

Yellow Stone Mining District, Pima County, February 21, 1880 (Recorded February 24, 1880, in Book 1 of Misc. Records, p. 68 and published March 4, 1880, in *The Weekly Nugget*)

New Mining District

At a miners' meeting held in Yellow Stone District, about eight miles north from Tres Alamos, on the 21st of February, 1880 for the purpose of forming a mining district, the following resolutions were adopted:

RESOLVED That the name of the district shall be Yellow Stone.

Second. That the United States mining laws shall be adopted as a whole.

Third. For the purpose of doing assesment work $6 per day shall be allowed for the purpose of holding claims.

Fourth. For the purpose of calling a miners' meeting, three notices shall be posted in three of the most conspicuous places in the district for ten days, and signed by not less than nine mine owners in the district.

Fifth. The records of Pima county shall be the bona fide records of the district.

Sixth. Five days shall be allowed for monumenting claims. Notices to be posted with the name of claim and owner on each monument.

Seventh. The district to be ten miles square, Warren's springs being the center of the district.

C. NOLL, Chairman
C. W. RUSSELL, Secretary
JOHN W. WATSON,
FRANK PARKER
GEO. KNUDSON,
S. P. OVERTON,
JAMES LIPPETT.

1880.04.10

Golden Basin Mining District, Mohave County, April 10, 1880 (Recorded April 19, 1880, in Book 1 of Misc. Records, pp. 489–90)

Golden Basin Mining District Laws

At a meeting of miners held at Cold Springs in Mohave County Arizona April 10th 1880, there was present the following named persons, R. G. Patterson, O. J. Kenncer, G. W. Mausan, John Fox, Charles Clark, J. S. Curtis.

R. G. Patterson was elected chairman of said meeting and J. S. Curtis, Secretary.

It was resolved that a district be formed under the name of the "Golden Basin" Mining District, within the following boundaries, to wit:

Beginning at the Colorado River and the Eastern side of Sacramento Valley, thence Easterly along Southern Bend of Colorado River about 20 miles to the junction of said river and the Western side of Wallapai Valley; thence southerly along the Western side of Wallapai Valley to the intersection of the line forming due East and West though Mountain Springs, being a distance of about 35 miles, thence along said line about 20 miles to Eastern side of Sacramento Valley. Thence about 35 miles to point of beginning.

It was further adopted that the mining laws of Congress be adopted as the rules and regulations of this District.

J. S. Curtis, Secretary, R. G. Patterson, Chairman of meeting.

Recorded at the request of R. G. Patterson, April 19th A.D. 1880, at 12 N.

Caldwell Wright
County Recorder

1880.06.05

Pajarita Mining District, Pima County, June 5, 1880 (Recorded June 15, 1880, in Book 2 of Misc. Records, pp. 212–16)

Miners Meeting

At a meeting if miners held this 5th day of June 1880 at McArthur's store, Pajarita mines, Pima Co A.T. On motion, Mr. J. M. McArthur was elected Chairman and J. B. Lowrey, Secretary. The Chairman stated that the meeting was called for the purpose of organizing a Mining District.

No. 1 It was moved and seconded that a Mining District be formed to be known at the Pajarita Mining District and laws enacted for the Governance of the same. Motion Carried.

2. It was moved & seconded that the district so formed embrace the following territory, to wit: Commencing at the point on the Santa Cruz River where the river intersects the boundary line between Arizona and the State of Sonora, Mexico, thence following down the river to the old mill at the town of Tubac, thence to the peak known as the Oro Blanco Picacho, thence South 20 West to the boundary line between the Territory of Arizona and the State of Sonora, Mexico, thence along the said line to the place of beginning. Motion Carried.

3. It was moved & seconded that the County Recorder of Pima Co, by virtue of his office shall be ex-officio Recorder of this District. Motion Carried. It was moved & seconded that the Mining Laws of the United States of America now in force, be adopted for the government of this district. Motion carried.

4. It was moved & seconded that in the location of mines in this district copies of the notice of location must be placed on the mines before any legal record of the same can be made by the Recorder. Any location not so made shall be null and void. Motion carried.

5. It was moved & seconded that all notices of location must be filed with the Recorder within sixty (60) days from the actual date of location. Motion Carried.

6. It was moved & seconded that the Country Recorder shall be entitled to a fee of two ($2.00) dollars for each and every notice recorded by him. Motion carried.

7. It was moved & seconded that the Records of Pima County are hereby adopted as the bona fide Records of this District. Motion carried.

8. It was moved & seconded that the annual meeting of the voters of this district shall be held on the first Monday in June of each year. Motion carried.

9. It was moved & seconded that ten (10) days prior to the date of holding the annual meeting the Recorder shall place or cause to be placed or posted in three of the most conspicuous places in said Distinct a notice stating the time when & place where said meeting shall be holden & shall designate in such notice that the meeting shall be holden for the purpose of transacting all & every kind of business which may property be brought before it. Motion carried.

10. It was moved & seconded that at each annual meeting the voters shall elect their Chairman & Secty, who shall hold office for one year or until their successor are elected. Motion carried.

11. It was moved & seconded that the Secty of each meeting shall keep full & complete records of the proceedings of their respective meeting & cause the same to be placed on record with the County recorder. Motion carried.

12. It was moved & seconded that these rules Regulations & By Laws shall not be altered or in any way changed except at a regular annual meeting of the miners of said district, and then only by a legal vote of two thirds of all the voters present & voting. Motion carried. It was moved & seconded that any & all persons who are citizens of the US of America or who have declared their intention to become such and own shares of stock or interests in any mine in the district or who have worked in any mine in the district for the twenty days preceding such meeting shall be considered a legal voter and entitled to vote at a miners meeting. Motion carried. It was moved & seconded that all mines hereafter located in this district shall be marked by end monuments or stakes at least 18 inches in height above the ground with sufficient marks placed in or upon them to show to which end of the claim they designate—If stakes are used they must be sunk at least six (6) inches in the ground & have a blaze and figures upon one side. Motion carried.

13. It was moved & seconded that the owner or owners of mines in this district shall be entitled to one mill site of five (5) acres for each claim or

mine held by him or them to be known & designated by the name of the mine for which it is located and to be recorded and located in the same manner as a mining claim. Motion carried.

14. It was moved & seconded that a special meeting may be called at any time to transact such business as may be brought before it by posting at least three (3) notices specifying the business for which the meeting is called in conspicuous places in the district signed by the chairman & secty & at least five (5) voters of the District; said notice to be so posted at least ten (10) days before the date of meeting & no business shall be transacted at such meeting other than that specified in the notice. Motion carried.

[16.] It was moved & seconded that these rules, regulations & by laws shall be recorded in the office of the County Recorder of Pima Co. & shall be in full force and effect from and after this 5th day of June 1880. Motion carried.

17. It was moved & seconded that the above proceedings and the proceedings of any subsequent meting shall be signed by the Secty & Chairman and transmitted by them to the County Recorder without unnecessary delay. Motion carried. It was moved & seconded that the meeting adjourn sine die. Motion carried.

J. M. McArthur,	Chairman
J. B. Lowrey,	Secty
W. E. Jervosse,	Miner
J. A. Rockfellow	"
A. K. Brewer	"
John Callore	"
T. B. Kerr	"
James Fine	"
M. A. Kerr	"
James Peterson	"
T. Beltran	"
Henry McGrady	"
Trinidad Farriori	"
Santaigo Beltran	"
W. D. Smith	"
Richard Thomas	"
Nick Simons	"
Jos Trewin	"

Filed and recorded at request of J. M. McArthur June 15th A.D. 1880 at 8 AM.
S. W. Carpenter, County Recorder.
By H. Hatch, Deputy

1880.12.06

Palmetto Mining District, Pima County, December 6, 1880 (Recorded December 15, 1880, in Book 2 of Misc. Records, pp. 41–42)

At a mass meeting of the miners of the Palmetto district in Pima Co. Arizona, it was resolved that the said Palmetto District be governed by the Mining laws of the United States and that no local laws be passed that would in any way conflict with the said laws of the United States.

Article (2) Be it further enacted that the said Palmetto Mining District of Pima Co. bounded as follows to wit: Commencing at a point on the Sonoita River at or near Sanfords House, thence down the Sonoita River to the junction of the Sonoita & Santa Cruz Rivers to the Sonora line, thence along the Sonora line to a pass between the Patagonia Mountains & the San AnnTone Mountains, thence along the summit of the Patagonia Mountains to the point on the Sonoita River about two Miles Easterly from the Old Astic Mill site, thence down the Sonoita River to place of beginning.

Article (3) Be it enacted that the Country Recorder be the Recorder for the Palmetto District.

Article (4) that the minutes of this meeting be published in the Arizona Daily Star.

Dated December 6th, 1880.

Secretary
A. W. Osten

President
G. W. Atkenson

Filed and recorded at request of James Spudy, Dec 15th A.D. 1880 at 2:30 P.M.
S. W. Carpenter, County Recorder
By _____, Deputy

1880.12.27

Minnesota Mining District, Mohave County, December 27, 1880 (Recorded January 29, 1881, in Book 1 of Misc. Records, pp. 514–15)

At a meeting held this day the 27th of December 1880 to form a district. C. C. Jones was chosen chairman and N. S. Lewis Secretary.

The following By Laws was adopted:

Art 1st This District be known by the name of Minnesota District and Bounded as follows: Commencing at Johnsons Rock on the Colorado River and running east 20 miles, thence North 20 miles, thence to Roaring Rapids, thence down the River to place of beginning.

Art 2nd There shall be a President and Recorder Elected once a year at the annual meeting on the Tenth day of January of each year.

Art 3rd All claims located in this district shall be located under the United States laws passed May 10th 1872 and as amended.

Art 4th All Locators shall have 20 days in which to Record and build monument.

Art 5th It shall be the duty of the Recorder to keep suitable Books for Records and open for inspection in all business hours.

Art 6 It shall be the duty of the President to Post Notices Thirty days prior to the election for President and Recorder. President shall act as Chairman at all miners meetings, in his absence one Pro-tem may be elected by the miners calling the meeting to serve without salary.

Art 7 The Recorder shall have three Dollars for Recording each claim he Records.

Art 8 A meeting may be called by Five claim holders by Posting three notices in the most conspicuous places in the said District thirty days before the date of meeting to amend said Bye laws.

Art 9 A man may locate one claim of Fifteen Hundred feet and no more on same ledge.

A True Transcript of the Bye Laws of Minnesota Mining District, Mohave County Territory of Arizona.

S. F. Walker, Recorder
December 27, 1880
Chairman C. C. Jones
Sec N. S. Lewis

Recorded at the request of S. F. Walker January 29th A.D. 1881 at 11 oclock AM

John K. MacKenzie
County Recorder

1881.01.01

Greaterville Mining District, Pima County, January 1, 1881 (Published March 17, 1881, in *Daily Citizen*)

Laws of Greaterville District, Pima County, A.T.

Section 1. This district shall be known as the Greaterville District.

Sec. 2. The boundaries of the Greaterville district shall be defined and established as follow, viz: Commencing at the point where the road from Thos. Gardiner's ranch joins the road coming down the Ophir Gulch from Greaterville; thence northerly to Mescal Springs; thence westerly to the northernmost point of the mine location known as the Catacomba, recorded in the records of the Smith District, folio 267; thence westerly to Alamos Ranch; thence south to T. Welisch's ranch; thence southeasterly to the site of Gardiner's "old" sawmill; thence easterly to the place of beginning, and shall include all the territory inclosed within those lines.

Sec. 3. The records heretofore kept and known as those of the Smith District, from the organization thereof on the 17th of March, 1875, to the date of the organization of this Greaterville District, are hereby declared accepted and established as the only and true valid district records heretofore kept relating to the territory of the said Smith District, as originally organized, and consequently also of the territory of this Greaterville District, and are to be upheld and respected as in force hereto fore, now and hereafter according; said records, so far as they relate to the territory inclosed in the Greaterville District, being acknowledged and adopted as the records of this district and the Recorder is hereby authorized and directed to keep possession and take care of them as such.

Sec. 4. No person shall be entitled to vote at any miners' meeting in this district unless he owns a claim (or an interest in one) in this district.

Sec. 5. The annual meeting of the voters of this district shall be held on the first Saturday in January of each year hereafter, at which meeting an election shall be held of Recorder for the District, and such other business transacted as may be found necessary. Ten days previous notice shall be given of said meeting by the record posting notice thereof in each of three prominent places in the district.

Sec. 6. No person shall be eligible to the office of Recorder unless he is a mine owner in and a resident of this district.

Sec. 7. The Recorder so elected shall hold his office for the term of one year from the date of his election and until his successor is duly elected and qualified.

Sec. 8. In case of death, resignation or removal of the Recorder of this district, any five legal voters may call a special meeting to elect a new Recorder, by posting notice thereof in each of the three prominent places in the district ten days prior to the date set for said meeting, and a Recorder may be elected by a majority of all the legal votes cast at said election.

Sec. 9. The duties of the Recorder shall be as follows: First—It shall be his duty to be present at every regular meeting and by virtue of his office to be and act as Secretary of such meeting. Second—On presentation of any notice of mining claim and location for record, it shall be his duty to examine such notice and to go in person or by deputy to the location and examine the same, and satisfy himself that it lies within the district; that the laws, rules and regulations of this district have been fully complied with therein and thereon, to refuse to record any such notice until such compliance has been effected or where such location conflicts with or trespasses on a previous location properly made and legally holding the ground. Third—it shall be his duty after satisfying himself that a notice of a claim and location is entitled to record, to receive the same, noting thereon the day and hour of its receipt and to record the same in the record book kept by him for that purpose, by making a fair and correct copy of said notice and the day and hour of its receipt by him, and to certify thereon that it has been so recorded and that his aforesaid examination has been performed. Fourth—It shall be his duty to carefully keep and record the minutes of all miners' meetings in a proper book to be kept by him, and to preserve from harm or loss all valuable papers and books entrusted to him relating to the district. Fifth—On written application to such effect from five legal voters of this district it shall be his duty to call a miners' meeting by posting notices in each of the three prominent places in this district at least ten days previous to the date set for such meeting.

Sec. 10. The fees of the Recorder shall be two dollars for examining locations, recoding and certify notice of same, and one dollar for any other certificate officially made by him.

Resolution adopted January 1, 1881.

—Resolved, that it is recommended that claim owners shall furnish their post office address to the Recorder, and in case the recorder learns of any thing out of order on their claim, he shall notify the owner by postal card.

Sec. 11. The size of any lode claim hereafter located in this district may equal, but not exceed, fifteen hundred feet in length in and along the lode, by three hundred feet in width on each side of the center of the lode; provided no prior legal claim exists within such limits.

Sec. 12. The size of any placer claim hereafter located in this district may equal, but not exceed, 1000 feet in length on the gulch, by 150 feet in width on each side of the center line of the gulch.

Sec. 13. The size of any claim and location made to secure and hold a spring, or a well dug by the locator furnishing water, may equal, but not exceed, 200 feet square, the sides facing respectively north, east, south and west.

Sec. 14. The boundaries of all claims and locations shall be defined by stakes set in the ground so as to show a height of at least four feet above ground, and they shall be braced by dirt or rock heaped up around the foot of each stake to a height of at least one foot. Each stake must be placed on the side facing the direction in which the lines run from it, and there shall be such a stake at each corner of the location, and one such at the center of each end line, and one such at a prominent place on each side line, as near the center of each line as possible and there shall also be placed at the discovery shaft a similar stake, on which shall be placed a notice stating the course and distance thence to the beginning boundary stake, which shall be the center stake of the end line nearest to such discovery shaft, beginning at such beginning stake with No. 1, the boundary stakes shall be numbered from one upwards with Roman numerals cut on the left hand place facing toward the succeeding number, the numbers invariably succeeding in order in the direction which is to the left of a person facing the claim; and the location notice shall be posted at the beginning stake, and shall give the courses and distances from stake to stake as nearly as possible, the end lines being always parallel to each other.

Sec. 15. The location notice shall contain the name of the lode, the name of the locator and date of the location, and the dimension of the claim, and a full description therein.

Sec. 16. No location is to be held as validly claimed until after the location notice is put upon on it at the prospecting point, and within ten days from the time said notice is put up at least nine stakes must be in place; and within thirty days after said notice is put up the location notice must be deposited for record with the Recorder of the District. And any failure to comply with any one of the requirements of this section shall render the claim void and subject to relocation by others, in case the original locator does not fulfill the above conditions prior to such relocation.

Sec. 17. The laws of the United States and Arizona now in force, and hereafter passed, shall rule this district in all questions which are within our right to govern, and which we have failed to mention or provide for herein.

Sec. 18. Fifteen legal voters of this district shall form a quorum to transact the business of a legally called miners' meeting of this district.

Sec. 19. In all cases of disputes regarding claims each party may choose one arbitrator, and the two so chosen shall choose a third. And such arbitrators shall endeavor to settle the dispute equitably.

Sec. 20. All laws and parts of laws heretofore governing the territory of this district which are in conflict herewith are hereby repealed.

Sec. 21. Sections 13, 16 and 22 were adopted January 1, 1881, to be in force on and after that day, and all the balance were adopted November 27, 1880, to be in force from and after that date.

Sec. 22. Resolved, That in working assessment labor shall be valued at four dollars per day of eight working hours.

Meeting of January 5, 1893

Special Meeting Mining District of Greaterville Arizona

January 5th 1893

E.N. Fish, Chairman

Resolved by the members present, That Jesus Quiros having at a dance held in the school House New Years day displayed a revolver and fired two shots from same within the building while the dance was in progress is a man who is an undesirable character in this community and that he be ordered to immediately leave this vicinity under penalty of being turned over to the officers of the law for their action. This to take effect within 24 hours from notice.

Committee appointed by the chairman to inform said Jesus Quiros of the action of the assembly.

The meeting then adjourned to meet January 21st 1893.

1881.01.01

Seattle Mining District, Apache County, January 1, 1881 (From BLM record)

The Miners in the Gila Range Hold a Meeting and Organize a New District

The miners of the Pueblo Viejo country, Apache county, met at the central point on Saturday, January 1, 1881, for the purpose of forming a new mining district. Capt. J. R. Rice was elected Chairman, and C. H. Snyder Secretary.

On motion the Chairman appointed Messrs Jones, Powers and Walker a committee to consider the subject of name and boundaries of the district, and Messrs Snyder, Scott and Young a committee to draught a series of by laws to govern the same. In a short time the Committee on Boundaries made the following report which was adopted:

The new district shall be known as the Seattle District, and shall be bounded as follows, to wit: Beginning at the mouth of the Rio Bonito, thence running west down the Rio Gila to Camp Thomas Military Reservation; thence running north to the Apache Indian reservation; thence running east along the line of the Indian Reservation to the Rio Bonito, thence south down the Rio Bonito to the place of beginning.

The committee on by-laws reported the following, which was adopted:

Article 1.—The district shall have a Recorder, to serve for the period of one year, and a regular district meeting shall be held annually on the first Monday in November for his election.

Art. 2—The said officer, at the expiration of his term of office, shall turn over the records and other papers in his possession to his successor in office

Art. 3—Three mine owners can at any time have the recorder to call a miner's meeting, and such meeting shall be called by posting notices in four conspicuous places in the district, and shall consist only of mine-owners.

Art. 4.—Ten mine-owners shall constitute a quorum to act in miner's meeting.

Art. 5.—The District Recorder shall receive two dollar for recording each mining location notice.

Art. 6.—All mines located in the district must be recorded in the office of the District Recorder within 30 days of location.

Art. 7.—That none of the by-laws of this district shall be changed unless a special meeting be called and governed by a two-thirds majority of mine owners.

On motion of Mr. Young the United States and Territorial Mining Laws were adapted as the laws of the district.

C. H. Snyder was elected Recorder of the district.

The Secretary of the meeting was instructed to forward a copy of its proceeding to the Arizona Citizen for publication.

C. H. Snyder
Secretary

J. C. Rice
Chairman

Resolution By Mr. Johnson, seconded by J. S. Lavorino, that all Mining Claims located in this Dist. After this date there shall be done at least 5 days worked to be completed within 60 days of location of said claim. This to apply to all claims lead lode or placer claims.

Resolution By Mr. Don Wendt, seconded by Mr. Young

That a ten (10) foot hole, or work equivalent to same, shall be done on each claim, located after this date within 60 days. This to apply to lead or lode claims.

1881.02.12

Weaver Mining District, Mohave County, February 12, 1881 (Recorded February 21, 1881, in Book 1 of Misc. Records, pp. 517–18)

Resolutions and By-Laws of Weaver Mining District

At a meeting of Miners of Weaver District held at the Weaver Mine, Mohave County, Arizona on February 12th, 1881, when the following Rules and Regulations were adopted.

This District shall be known as Weaver District and bounded as follows.

Commencing at the Southeast corner of Minnesota District at and to a spring South on said mountain down to a point known as Round Island on the Colorado River, thence up the River to a point known as Johnstons Rock, thence east to the starting point.

And be it further resolved that no Recorder shall be elected for this District and all claims located in this District shall be Recorded in the County Records within Sixty days from date of location.

And be it further resolved that these laws with the Laws of Congress and of this Territory are deemed sufficient for the government of this District.

On motion adjourned Sine Die

B. N. Coleman, Chairman
John H Weaver, Secretary

Recorded at the request of John H. Weaver February 21st A.D. 1881 at 45 minutes past 4 o'clock P.M.

John K. MacKenzie
County Recorder

1881.03.12

Wrightson Mining District, Pima County (Published in *The Arizona Citizen*, Mar. 22, 1881, and in GLO Microfilm)

General Call of the Miners of Wrightson District.

Santa Rita Mountains, March 12, 1881

Whereas, in the absence of the Chairman we, the undersigned miners of this district, have elected Mr. L. W. Wakefield as temporary Chairman.

A motion was made and passed that the office of District Recorder should be, and is, abolished in Wrightson District, and the recording from the date is to be done only in the office of the County Recorder.

A motion was made and passed that the law compelling the sum of $25 worth of work to be done inside of ninety days is also abolished from this date.

A resolution was also passed requesting the Secretary of this meeting to forward a copy of the same to the publishers of the Tucson Citizen, asking him to do us the favor of publishing the same in his paper.

A resolution passed that this meeting is adjourned sine die.

L. W. Wakefield, Temporary Chairman
S. L. Parke, Secretary
E. R. Devoe
C. Kramer
H. Brodt
C. Coonte

1882.04.06

Lost Basin Mining District, April 6, 1882, Mohave County (Recorded April 26, 1882, in Book 2 of Misc. Records, p. 701)

Bye Laws of Lost Basin Mining District

Lost Basin Mining District

At a meeting held by the miners of the above named district it was resolved that it should be known as Lost Basin Mining District, situate in the County of Mohave, Territory of Arizona, and more fully described as follows to wit:

Commencing at the centre of Wallapai Wash where said wash comes into the Colorado River and running from thence or nearly so Easterly along the South bank of the Colorado River to Pierce's Ferry on said River, thence up along the main wash and Wagon Road in a Southerly direction leading from Pierces Ferry to Grass Springs, thence westerly to Pattersons Well (on the Wagon Road leading from Grass Springs to the Eldorado Mine), thence further West to the Eastern boundary of Gold Basin District and from thence Northerly along the Eastern boundary of Gold Basin District to centre of Wallapai Wash and place of beginning.

Be it resolved that no District Recorder shall be elected, that the county Recorder shall be Recorder for this District.

Be it further resolved that all locations made prior to the formation of this District and within the boundary lines of said District shall come under the action of rules and regulation adopted at this meeting.

Resolved. That the United States Mining law of May tenth (10th) 1872 and the amendments thereunder shall govern this District.

The foregoing resolutions were adopted this Sixth day of April, 1882.

H. S. Devine — James Hodgon
Secretary — Chairman

Recorded at the request of the Miners of Lost Basin Mining District April 26th A.D. 1882 at 217 p.m.

1882.07.01

Copper Mining District, Maricopa County, July 1, 1882 (Recorded in Book 1, Misc. Records, pp. 349–53)

By virtue of a Notice duly signed and posted on the 20 day of June 1882, a meeting of miners was held at the residence of Dr. Jns L. Gregg as Chairman and J. J. Hill as Secretary. The following are the laws passed to govern the District.

Sec. 1. The Mining District shall be called Copper Mining District.

Sec. 2. The District shall embrace the following described and bounded Territory, commencing at a point when the Section line between Section 14 and 15 in T.1 North of R.4 East crosses the main channel of Salt River thence south on said line to the south line of Maricopa County; thence west on said Maricopa County line to the main channel of the Gila River; thence down said Gila River to the junction with Salt River; thence up the main channel of said Salt River to the place of beginning.

Sec. 3. The County Recorder of Maricopa County by virtue of his office, shall be ex-officio Recorder of this District.

Sec. 4. Chapter 6 of title 32 Revised Statutes of the United States Mining Laws is adopted as this article.

Sec. 5. In the location of mines in this District, copies of the notices of location must be placed on the mines before any legal Record of the same can be made by the Recorder. Any location not so made shall be null and void.

Sec. 6. All location notices must be filed in the office of the Recorder within sixty days after the actual date of location.

Sec. 7. Records of Maricopa County are hereby adopted as the bona fide Records of this District.

Sec. 8. The annual meetings of the voters of this District shall take place on the first Monday in July of each year.

Sec. 9. Ten days prior to the annual meeting the Secretary shall give notice by causing at least three notices to be posted in the District giving notices when and where the meeting will be held, and for what purpose.

Sec. 10. At each annual meeting, the voters of the District shall elect their Chairman and Secretary who shall hold office for one year or till their successors are elected and qualify.

Sec. 11. The Chairman and Secretary of this meeting shall hold office from this first day of July for one year or until their successors are elected and qualify.

Sec. 12. The Secretary of each meeting shall keep full and complete records of the meetings and proceedings of their respective meetings, and cause the same to be placed on Record in the office of the County Recorder.

Sec. 13. These rules, regulations and by-laws shall not be altered or in anyway changed except at a regular annual meeting of the miners of said District, and then only by a legal vote of two thirds of all the voters present and voting.

Sec. 14. Any and all persons who are citizens of the United States of America, or who have declared their intentions to become such and own shares of stock or interests in any mine in the District, or who has worked in any mine in the District for the 20 days preceding such meeting shall be considered a legal voter, and entitled to vote at a meeting.

Sec. 15. Five dollars per day shall be allowed for each and every eight hours work performed upon a miner for the purpose of holding title or performing the necessary amount of work for a patent and no other expenses shall be considered as expended for the purpose of holding or perfecting title.

Sec. 16. All mines hereafter located in this District shall be marked by end and corner monuments, on stakes at least eighteen inches in height above the ground with sufficient marks and figures placed in or upon it to show which end or corner of the claim they designate. If stakes are used they must be sunk at least six inches in the ground and have a blaze and figures upon one side.

Sec. 17. All locations made and recorded previous to the adoption of these rules, regulations and by-laws are hereby legalized. If made in conformity with the United States Mining Laws.

Sec. 18. These rules, regulations and by-laws shall be filed and recorded in the office of the County Recorder of Maricopa County, Arizona, and shall be in full force and effect after this the first day of July A.D. 1882.

Sec. 19. The above proceedings and the proceedings of any subsequent meeting shall be signed by the Chairman and Secretary and transmitted by the Chairman and Secretary to the County Recorder without delay.

Jns L. Gregg, Chairman

I hereby certify that the above is a correct statement of the proceedings and of the laws adopted for the Copper Mining District, in Maricopa County, Arizona, this first day of July A.D. 1882.

John J. Hill, Secretary

Recorded at request of J. J. Hill July 6 A.D. 1882 at 9:30 o'clock A.M.

R. F. Kirkland, County Recorder
By E. B. Kirkland, Deputy

1882.08.10

Little Cottonwood Mining District, Mohave County, August 10, 1882 (Recorded August 17, 1882, in Book 2 of Misc. Records, pp. 718–19)

Little Cottonwood Mining District, Mohave County, Arizona

In response to a call of the miners prospecting in his vicinity for the purpose of forming a Mining District, the following named persons assembled near the Lower cabin on Little Cottonwood Creek, on Thursday, August 10th 1882, due notice of which was posted ten days previous. J. W. Grant, Wm Krempe, A. H. Hanson, J. B. Adams, Danl Garnes, Dan'l B. Leary, Jas Noonan, Geo Wald, Henry McAlister, H. Stout, Ike Bould, W. W. Alden, J. W. Michael, Geo W. Brooks, Wm Chaffin.

The following Bye Laws were adopted and J. W. Grant elected chairman for the District and J. B. Adams Secretary for the term of one year.

Bye Laws

Article I This District shall be known and called Little Cottonwood Mining District.

Article II The Officers of this District shall consist of a chairman and Secretary who shall hold their offices for one year from the date or until their successors are elected.

Article III The chairman shall have power to call a meeting of the miners at any time when in his judgment it will be of interest to the camp by posting there notices giving at least five days notice and it shall be his duty to call such meeting on the written request of ten (10) mine owners of the District.

Article IV. The Chairman shall preside at all meetings of the District or in his absence the secretary shall act in his place, or in the absence of both the meeting shall elect both Secretary and chairman pro tem.

Article V The centre of this District shall be the falls on Little Cottonwood Creek and extend from the falls five miles north, five miles South, five miles East and five miles West.

Article VI. The amount of five ($5.00) dollars shall be considered and allowed as expended for each days assessment work performed on each and every claim.

Article VII. The by-Laws can be changed and amended at any regular meeting of the miners by a two thirds vote of the miners present.

J. B. Adams Sec J. W. Grant Pres

Recorded at the request of J. B. Adams August 17th A.D. 1882 at 3:10 P.M.

John P. MacKenzie, County Recorder

1882.11.18

Richardson Mining District, Mohave County, November 18, 1882 (Recorded in Book 1 of Misc. Records, pp. 614–18)

By Laws of Richardson Mining District

Pursuant to a Public notice, duly posted in Little Cottonwoods Mining District, Mohave County, A.T. a Miners Meeting was convened on the 18th day of November for the purpose of filling vacancies and electing officers for the last quarter of 1882 and 1883, and to provide measures for the local government of this Mining District.

Meeting convened at 2.20 p.m. Present Joseph A. Ashlock, Frank Duffee, Geo. Waltz, Dan'l Remington, James Fulton, Sam'l L. Bernie, Harry Moore, Henry P. Bennett, H. Stout, Jms. Basshart, Fred Bauer and J. E. Stevens.

House called to order at 4 o'clock. Frank Duffee called to the chair, and H. P. Bennett announced Sec Pro Tem.

On motion of J. A. Ashlock, Frank Duffee was duly elected to the office of Chairman for the last quarter of 1882 and 1883.

On motion of D. Remington, H. P. Bennett was duly elected Secretary for the same length of time.

On motion of H. P. Bennett, Sam'l L. Bernie was unanimously elected District Recorder for the last quarter of 1882 and 1883.

On motion of S. L. Bernie, J. A. Ashlock, A. Remington and J. Fulton were appointed as committee to draft By-Laws for the regulation of the Mining Camp, and ordered to report on Saturday, Nov. 25th for ratification.

On motion of D. Remington, the meeting adjourned until Saturday Nov. 25th 1882 at 4 p.m.

H. P. Bennett
Sec.

Little Cottonwood Mining District

November 25th 1882

Miners convened at the cabin of Frank Duffee pursuant to adjournment. Miners present in full force, Frank Duffee presiding.

House being called to order by the Chairman who stated the object of the meeting, he also stated that in consequence of the apathy of the meeting locators of an earlier day, the camp was in a backward State, that but little work had been done consequently its merits were as yet but little understood; but when labor had been performed upon property, with a view to its development, now new district could be cited to in the whole area of his broad experience as a mining man that would compete with this camp. He also complained of the "misnomer" of the camp as tending to create confusion and lead mining men taken on account of Little Cottonwoods Mining Districts being too numerous over the country and necessitating in his mind a reorganization seemed necessary and giving the District a name that shall acquire that fame in Arizona mining history that will admit of no second standing by any other Mining District in the Union, as the present outlook fully justifies, after which remarks the Chairman resumed his seat, declaring the meeting open for the transaction of business.

On motion of S. L. Bernie, seconded by J. A. Ashlock, the Little Cottonwood Mining District was changed to "Richardson" Mining District, Mohave Co. A.T. being carried by a unanimous vote.

Under the head of reports of committees, J. A. Ashlock, chairman of Committee for By-Laws, presented the following

By-Laws

Resolved: That the Local Recorders fees shall be three Dollars ($3.00) for each certificate placed upon the Recorders Books of the district.

Resolved: That all mining claims located within this the "Richardson" Mining District shall be duly recorded within the District at the local office for the general information to newly arrived mining prospectors and to avoid any complex questions that might arise from any prior locations made by other mining prospectors.

Resolved: That the government tax or assessment labor expended upon any lode claim located within this Mining District shall be done or caused to be done by the locators as follows, to wit: Labor per diem in which blasting becomes necessary on a prospect shall be valued at $8.00 per day; and labor performed in which no blasting shall be required, $5.00 per day, length of time in which to perform assessment, to conform to the U.S. Statutes.

Resolved: That it shall be the duty of the local Recorder to be present and superintend the surveying and staking of all Lode Claims with this District, and for every lode claim so located the recorder shall receive as his fees the sum of $5.00.

Resolved: That in case of any dispute upon any claim that might arise out of any location made on any lode claim in this Mining District upon complaint being made to the local recorder, he shall have power, and is hereby authorized to appoint two disinterested miners within this Mining District to proceed to an investigation, and with the assistance of these two miners shall have power to decide upon the validity of the adverse claims of the respective parties, and the decision of the recorder in such cases shall be final. In all cases of disputed claim titles the recorder's fees shall be $5.00 for each day or fraction of a day so expended in the interests of such parties; such fee to be paid by the party against whom judgment shall be rendered. That all complaints of whatsoever nature or kind, arising out of mining claims by and through any party or parties present or that may hereafter become interested in Mining Claims within this Mining District shall be adjusted by the recorder or the recorder and two disinterested miners duly authorized by said recorder of this Mining District.

Resolved: That the meets and bounds of the Richardson Mining District shall be and are hereby declared as follows:

Beginning at a monument of stone and a stake erected on Mount Bennett, and marked "Mt. Bennett" distance about Six Miles East of Hackberry or nearly so, and running west to the Railroad Grade, thence South four (4) miles; thence East ten (10) miles, thence north along the divide of the slate, range eight (8) miles; thence West ten (10) miles to the Railroad Grade, thence South to the point of beginning.

Officers

Resolved: That the officers of this organization shall consist of a local recorder, chairman and secretary; those officers to be elected by vote annually, said annual election to be holden on the 1st day of January of each year after January 1883.

Resolved: That the duties of these several officers shall be as follows.

The Recorder shall call all special meetings for the purpose of settling any disputes that may arise among miners within this District, in regard to adverse claims & etc., he shall record all locations made within the district on the local record and keep the same open for the benefit of mining prospectors.

The Chairman shall preside at all regular and called meetings except at such meetings as are called by the recorder for arbitration.

It shall be the duty of the secretary to be present at all meetings and to record all the transactions connected therewith.

J. A. Ashlock
Jas. Fulton
D. Remington

Moved by Geo. Waltz that the By-Laws be accepted, and the Committee discharged, carried.

Moved by J.A. Ashlock that a copy of the proceedings be forwarded to the Office of the Mining News for publication, carried.

Moved by D. Remington that the proceeds of the Miners of Richardson Mining District be recorded in the office of the County Recorder, carried.

On motion meeting adjourned, Sine die

Frank Duffee — H. P. Bennett
Chairman — Secretary

Recorded at the request of H. P. Bennett December 28th A.D. 1882 at 4:30 P.M.

John K. MacKenzie
County Recorder

1890.05.24

Silver Hill Mining District, Pima County, May 24, 1890 (Recorded May 29, 1890, in Book 4 of Misc. Records, pp. 393–94)

Pima County Arizona Territory—At a meeting of a majority of the miners of Cerro Colorado and Silver Hill and vicinity at the Liberty mine 24th day of May 1890, Silver Hill District was organized with the following boundaries. Commencing at the S.E. corner of Township 20 range 10 S. Gila and Salt River Meridian—thence west six miles to SW corner of said township, Thence N. twelve (12) miles to adjoining Township corner, thence S twelve (12) miles to place of beginning. The following rules and regulations were unanimously adopted for the government of said District—1st All persons making a mining location in said Silver Hill District shall perform or cause to be performed at least five (5) feet of work so as to define the vein clearly, before putting the same of record, the forgoing to take effect from date of recording this instrument. The following officers were elected for the ensuing year.

Chairman W. Clark
Secretary Jos Robinson

After which the meeting adjourned subject to the call of the chairman

Joseph Robinson, Wm Clark, A. J. Bowman, Y. B. Contins, Wm. J. Clark, J. T. Bowman, T. S. Moore, Jas Larson, M. M. Clarke

Filed and recorded at request Henry Manger, May 29th 1890, at 4:20 p.m.

Jas. Shibell
County Recorder

1892.06.28

Indian Secret Mining District, Mohave County, June 28, 1892 (Recorded on July 2, 1892, in Book 2 of Misc. Records, p. 497)

The Southern boundary to Commence at Louis Seabright house known as "Mountain Spring" thence in an Easterly direction to Dolan's Spring, thence in a northerly direction to the Centre of Salt Spring Wash to the bank of the Colorado River thence along the bank of the river to the mouth of the Sacramento Wash, thence in a Southerly direction along the wagon road to Louis Seabrights, the place of beginning. June 28th 1892—

Caldwell Wright, Sec'y
Henry Schaefer, Chairman

Recorded at the Request of Caldwell Wright July 2d A.D. 1892 at 30 minutes past 8 o'clock A.M.

E. J. Godman
County Recorder

1893.12.30

Superstition Mining District, Pinal and Maricopa Counties, December 30, 1893 (Recorded in Book 4 of Misc. Records, pp. 533–38)

Minutes of Mining Meeting held at Goldfield Superstition District Pinal County, Arizona Territory, Dec. 30th 1893.

Meeting called to order as per Notice at the Jones house 7 O'clock P.M. W. A. Kimball acting as Temporary Chairman, J. H. Possenny, Temporary Secy.

A motion to adjourn to Mannouth Co. Boarding house for better communication carried.

Reassembled at Co. Boarding house at 7:30 P.M. Motion of the Chair to appoint a Committee of (5) Five to suggest names for officers for permanent organization carried. Chair appointed J. H Ramsey, J. F. Lewis, Jas Hawks, Wm Boliths and E. A. Robinson as such Committee.

Motion that the chair appoint a Committee of (5) Five for Boundaries and of Bylaws Carried. Chair appointed J. R. Morse, Chas Edwards, Samuel Grayson, M. L. Spear, Jas Coswell as such Committee.

Motion of recess for 30 minutes report Carried.

Reconvening at 8:30 Report of Committee. We your Committee for the organizational Bylaws submit the following, Lawrence Kimball President, W. E. Possenny, Vice President, J. G. Peterson, Recording Secretary: J. H. Possenny—chairman. Report Adopted unanimously.

Lawrence Kimball was escorted to the chair.

In absence of J. G. Peterson, J. H. Possenny was elected recording Secretary Pro tem.

Committee on Boundaries and Bylaws reported as follows:

Boundary lines to commence at the Junction of the Verde and Salt Rivers and thence South to the base lines Gila and Salt River Meridian, thence

East to the dividing Ridge between Marlins Ranch and the Fragien Canon, Thence North to Salt River, Thence down Salt River to place of beginning.

2d. This District Shall be known as the Superstition Mining District. The Laws to be recorded in Maricopa and Pinal Counties on or before Jan 15, 1894.

3d. The Officers of this District shall consist of President, Vice President and Recording Secretary.

4th. The duties of the President shall be to call Miners meeting at the request of (10) ten or more miners for the purpose of Changing the Laws and Regulations of this Mining District, to call said meetings by giving a notice not less than 30 days to be published in the Nearest News Paper and by posting notices in three Conspicuous places in the Town of Goldfield the object of said Meeting shall be set forth in said Notice.

5th. The duties of the Vice President shall be to act and perform all duties of the President in his absence.

6th. Duties of Recording Secretary shall be to keep a true record of proceedings of all Meetings of the Dist also record all laws, rules and regulations passed by such meetings and see that they are recorded in their respective Counties as heretofore provided.

7th. The regular meetings shall be held semiannually on the first Monday of June and the first Monday in December in each year.

8th. All persons [claiming] mining claims or interested in them or Miners or residents of the District shall be entitled to a voice in the Meetings [and the name of others].

9th. A Code of laws as follows—

Section 1st.

Dimensions of Locations. Claims on any leade, lode or vein bearing Gold, Silver, Cinnabar, Lead or other valuable mineral deposit shall consist of not more than (1500) fifteen hundred feet along the leade, lode or ledge and not more than (300) thee hundred feet on each side.

Section 2d.

Requites of Location. The discover of a location shall within (60) Sixty days from date of discovery Record his claim in the office of the Recorder in the County in which such Lode is situated by a Location Certificate which shall contain:

First, The name of the lode; Second, The name of the Locator or Locators; Third, the date of Location; Fourth, The n[umber] feet claimed on each side of the center of Discovery shaft; Fifth: The general course of Lode or near as possible.

Section 3rd.

Discovery Shaft and Staking. Before filing such Location Certificate the discoverer shall locate his claim by, 1st sinking a discovery shaft upon the Lode to the depth of at least (10) Ten feet from the lowest part of the [surface] of such shaft at the surface or deeper if necessary to show a well defined [vein]. Second, By posting at the point of discovery on the surface a plat or Notice containing the name of the Lode, The name of the Locator or Locators and the date of discovery. Third, By marking the surface Boundary of claims.

Section 4th.

Boundary Stakes. Such surface Boundaries shall be made by (8) eight substantial posts or Monuments of stone not less than (3) Three feet high, placed on at each corner one at center of each side and end line.

Section 5th.

Open cuts and Tunnel Discoveries. Any open cut, Cross cut or Tunnel which shall cut a Lode at the depth of (4) four or more feet below the surface of an adit exploring the vein at least (10) Ten feet below the surface along the lode from the point which the Lode may be in any manner discovered shall be equivalent to a discovery shaft.

Section 6th.

Sixty days for sinking a shaft. The discoverer shall have (60) days from the time of assuming or disclosing a lode to sink a discovery shaft thereon.

Section 7th.

Right of Way. All mining claims [previously] located on that may be hereafter located shall be subject to the right of way of any ditch, flume or piped lines for mining purpose or any tramway or Pack Trail [for mining] use or hereafter laid out across any such location. The users of any line, Ditch, Tramway or trail shall be liable for any damage done in the Construction or operation of such pipe line, Ditch, flume tramway or Trail.

Section 8th.

Relocation of his own claim by the Owner. If at any time the locator of any mining claim heretofore or hereafter located or his assigns shall be apprehend that his original certificate was defected erroneous or that the requirements of the laws had not been complied with before filling or shall be desirous of changing his surface Boundaries or taking in any part of an overlapping claim which has been abandoned such locator or his assign may file an additional Certificate, Providing That such relocation does not interfere with the existing rights of others at the time of such relocation and no such relocation or other record of things shall preclude the Claimant or Claimants from proving any title or titles as he or they may have held under previous locations.

Section 9th.

Relocation of Abandoned Claims. The relocation of abandoned lode claims shall be by sinking a new discovery shaft and fixing new boundaries in the same manner as the location of a new Claims or the relocator may sink the original discovery Shaft (10) Ten feet deeper than it was at the time of abandonment and renew or adopt the old boundaries so assuming the posts or Monuments if [missing] or destroyed. In either case, whether in whole or part, if the abandoned claim is taken, the location Certificate shall state that the whole or any part of the new location is located as abandoned property.

Section 10th.

Work to be done on location prior to December 30th 1893. Any claimant or claimants meeting on claiming to use Lode claims located prior to December 30th 1893 that have not complied with Section (3) three of these articles shall sink a shaft the depth of (10) ten feet on such claims before the first day of May 1894, or such claim shall be subject to relocation.

Report of Committee and adopted unanimously.

Motion to adjourn Sine Die Carried.

J. H. Possenny
Recording Secretary
Protem
L. P. Kimball
President

Filed and recorded at request of Chas. L Hall, Jan 8th A.D. 1894, at 9-A.M.

Jose M. Ochoa
Recorder

1896.04.04

Warm Spring Mining District, Coconino County, April 4, 1896 (Recorded in Book 1 of Promiscuous Records, pp. 89–90, Coconino County)

Minutes of Meeting of Miners, held for the purpose of a Mining District.

At a meeting of the miners and mine owners held on the ground occupied by mining claims lately located, held for the purpose of organizing a mining district, the following action was taken:

On motion of J. F. Burnes, C. F. Cox was elected chairman and secretary of the meeting.

On motion of J. F. Burnes, C. F. Cox was elected permanent Recorder of the District.

Moved and carried that the Township be not less than six miles square. Moved and carried that this be called the Davis Tp. and Warm Spring Mining District.

Moved and carried that a fee of One Dollar be allowed for recording each Notice of Location.

Moved and carried that a committee of three be appointed to draft By-Laws for the District, L. S. Duham, Joseph Millett, Jr., and J. F. Burnes were appointed as such committee and instructed to report Apr. 6 at 9 o'c A.M.

Moved and carried that we adjourned until Apr. 6, at 9 A.M. April 6, 1896.

Pursuant to adjournment meeting convened.

Report of Committee on By-Laws being called for and read, the By-Laws were taken up section at a time and acted upon and were adopted as follows:

By-Laws of Warm spring Mining District, drafted this 4 day of Apr. 1896.

Sec I. Be it enacted by the members of the organized mining district of Warm Spring, That, every individual locating mining claims in this district shall be required to have his claim recorded within thirty (30) days, and in case of failure to comply with this Sec. I, his claim shall be forfeit and subject to relocation.

Sec. II. Every person complying with Sec. I shall be allowed to claim all timber, water and precious metals within the boundaries of his claim, if mentioned in his Notice of Location.

Sec. III. That every Locator be required to do fifty (50) Dollars worth of work on his claim within six months from date of location; Said work to consist of a hold on the vein or lode of four feet square by five feet in depth.

Sec. IV. That no Locator shall take more than one claim outside of his claim of Discovery, on the same vein; and any other prospector shall locate but one consecutive claim on the same lead or vein.

(L. S. Dunham
Com. (Joseph Millett, Jr.
(J. F. Burnes

Members present and taking part in the meeting.

J. M. Davis
J. F. Burnes
J. P. Nielson
Joseph Millett, Jr.
Byron G. Millett
Azro DeMill
L. S. Dunham
Thomas P. Jensen
C. F. Cox, sec

Moved and carried that we adjourn sine die.

C. F. Cox.
Chairman & Sec.

Recorded May 16th, 1896, at five o'clock, P.M.

C. A. Bush, County Recorder

1901.04.18

Reed Mining District, Pima County, April 18, 1901
(Recorded in Book 6 of Misc. Records, pp. 380–81)

Whereas the miners resident in the vicinity of that mountain or peak known as Caley's Peak are desirous of forming a mining district, and whereas in pursuance of such desire the said miners have been called together to discuss and set upon the matters, and whereas a meeting of the said miners was called for the 18th day of April, 1901, at Donovan's Camp the said Caley's Peak, and whereas the undersigned miners were present at the said meeting. Now, therefore, we the undersigned miners residing near or having mining interests near the said Caley's Peak do hereby certify that at the said meeting it was unanimously decided that a new mining district should be formed including all that territory lying in and around the said Caley's Peak, and the following resolution was unanimously passed. Resolved, that the said mining district should hereafter be known and described as the Reed Mining District, and the boundaries of the said Reed Mining District are as follows:

To wit: commencing at a mound of stones five miles due North of said Caley's Peak, run thence East five miles to a monument of stones, being the N.E. corner of said District; run thence at right angles due South 10 miles to a monument of stones, being the S.E. corner of said District; run thence at right angles due West 10 miles to a monument of stones being the S.W. corner of said District; run thence due North at right angles 10 miles to a monument of stone, being the N.W. corner of said District; run thence at right angles due East 5 miles to the place of beginning.

That the said Caley's Peak is approximately in T12S, 4, R17E, G&SR Meridian on the South East slope of the Santa Catalina Mountains and about 2–1/2 miles West from the Sierra Palon; that the said Mining District and all thereof is situate in the County of Pima, Territory of Arizona. That at the said meeting W. T. Donovan, a resident of said Mining District, County and Territory, was duly elected the District Recorder of the said Mining District, to hold said office for the term of two years, and until his successor shall be elected and shall enter upon the duties of his office; that the said Mining District and all mining matters therein shall be governed and regulated by the mining laws, rules and regulations of the Territory of Arizona and the United States of America.

In witness whereof, we have hereunto set our hands, this 18th day of April, 1901.

P. P. Sullivan
Frank Foran
F. W. Jordan
Chas. W. Reed
John Caley
W. T. Donovan

Filed and recorded at request of F. H. Hereford May 7, 1901, at 11:00 a.m.
Chas. A. Shibell County Recorder

1905.01.09

Greenfield Mining District, Pima County, January 9, 1905 (Recorded on January 9, 1905, in Book 7 of Misc. Records, pp. 358–58)

To whom it may concern—We, the undersigned, being the majority of the miners within the district hereinafter described, hereby give notice, that at a meeting of the miners within said district upon the 2nd day of January 1905 held, the following resolutions were passed and duly adopted, to wit: Whereas, that certain district described as follows, to wit: beginning at a place called the Fresnal Ranch, in the County of Pima, Territory of

Arizona, in the eastern portion of the Comobabi Mountains, thence in a northerly direction a distance of about seven to ten miles to the 3rd Standard South Parallel; thence westerly along said Parallel a distance of about fifteen miles; thence southerly a distance of about twelve to fifteen miles to a point or place called Artesa; thence southeasterly a distance of about three to five miles to a place or point called Cholla; thence northeasterly a distance of about ten to twelve miles to the place of beginning—as appears by the records of the County recorder of said Pima County, has no definite or fixed name by which it may be termed or known, and, Whereas, said district so described hereinabove is becoming of importance, is being developed, and capital therein invested, therefore be it Resolved, That said district, so described as herein before set forth, shall be, and hereby is named and shall be known as and called "Greenfield Mining District of the Comobabi Mountains, Pima County, Territory of Arizona."

Witness our hands this 9th day of January A.D. 1915

N. P. Greenfield
W. B. Heslin
W. Warford

Filed & recorded request of N. P. Greenfield, Jan 9, 1905 at 3:15 P.M.

Chas. Shebell
County Recorder

1913.07.14

Purtyman Mining District, Coconino County, July 14, 1913 (Recorded July 18, 1913, in Book 2 of Promiscuous Records, p. 102)

Sedona July 14th 1913

We the undersigned Citizens of the United States, County of Coconino, State of Arizona desire a Mining District recorded and have named said mining district The Purtyman Mining District. Center of the District to be at the Old Purtyman Ranch above the Falls on Oak Creek and extend Fifteen Miles in any and all directions.

J. F. Demick
John J. Thompson
Joseph L. Lay
J. E. Purtyman

State of Arizona }
County of Coconino }

Before me, Dan J. Cronin, County Recorder in and for the County of Coconino, State of Arizona, on this day personally J. F. Demick, John J. Thompson, Joe L. Lay and J. E. Purtyman known to me to be the persons whose names are subscribed to the forgoing instrument and acknowledge to me that they executed the same for the purpose and consideration therein expressed.

Given under my hand and seal of office, this 18th day of July 1913.

Dan J. Cronin, County Recorder

Recorded at request of J. F. Demick, July 18, A.D. 1913 at 1 Oclock P.M.

Dan J. Cronin, County Recorder

APPENDIX C

Names

Names in Text

Aiton
Ajo
Akins
Alamos
Alcaraz, Diego del
Aldrich, Mark
Alejo, Rafael
Alfonso XI
Allen, Robert
Allyn, Joseph Pratt
Altar
Alto Arizona
Anza, Juan Bautista de
Apacharia
Apache Pass
Aranea Mine
Arevaca
Aribac (Arivaca)
Arissona (Aruzema)
Arivaca
Arizona Department of Mineral Resources
Arizona County
Arizona
Arizona Miner
Arizona Mining and Trading Company
Arizona Weekly Miner
Arizpe
Armijo, Manual
Arrivacca Ranche
Asiatics
Atahulpa
Audiencia
Aurora Mine
Azamor
Aztec Mine
Aztecs

Babispe
Baca Float No. 3
Baker, Elihu
Ball, Samuel
Bancroft, Herbert
Basela, Marriano
Bashford, Coles
Battle of San Jacinto
Bear Flag War
Beatty, William H.
Beers, Herbert P.
Benedict, A. C.
Benedict, Albert C.
Benedict, Kirby
Benitillas Mine
Big Bug
Big Sandy River
Bill Williams Fork
Blake, William P.
Blanding, William
Bocoachi

Encarnation Claim
Encinas, Marcilimo
Escalante, Juan Bautista
Escalante, Silvestre Velez de
Espejo, Antonio de
Estill, N. K.
Eugenie Quicksilver Mine
Eureka, Nevada
Ewing, Thomas
Extralateral Right

Fannin Massacre
Farfan de los Godos, Marcos
Fedirico, Jose
Fletcher, Frank
Fillmore, Millard
Fish, E. N.
Fisher, A.
Fitch, F. G.
Fleury, Henry W.
Fort Breckenridge
Fort Buchanan
Fort Rock Mining District
France
Freeman District
Freeman, Tho.
Frémont, John Charles
French Mill Site
French Mine
Fronteras
Fuller, George W.

Gadsden Purchase
Gamboa, Francisco Xavier I
Gandasa, Manuel
Garcia, José de Jesus
Gila City
Globe
Globe Mining District
Godroz, W. J.
Gold Mountain Mining District
Goldman, Jacob
Goodwin, John N.
Grass Valley Mining District
Gray, A. B.
Gray, Andrew B.
Greaterville Mining District
Green, T. B.
Greenlee Gold Mining District
Griegerick, Valentine
Groom, Robert W.
Grounds for Agriculture
Guadalajara
Guevavi
Gwin, William M.

Hamilton, Patrick
Hardy, W. H.
Hardyville
Harqua Hala
Harshaw Mining District
Hartley, William
Hassayampa Mining District
Hassayampa River
Hawley Mine
Hawley, Giles
Hayden, Charles Turnbull
Hayes, Rutherford
Heintzelman Mine
Heintzelman, Samuel N.
Herbert, P. T.
Heuvari
Higgins, W. S.
Hodge, H. C.
Hopi (Hopiland)
Hopkins, Charles
Horcasitas (Pitic)
Houston, Sam
Howell Code
Howell, William T.
Hualapai Mining District

Para, Lorenzo
Parral
Paso Verde
Patagonia Mine
Patagonia Mountains
Pattie, James O.
Pedis Possessio
Peeples, Abraham Harlow
Peplow, Ed
Philip II
Pimo Station
Pine Grove Mining District
Pinto Creek Copper Deposit
Pioneer Mining District
Pizarro, Francisco
Planchas de Plata
Plomosa Mining District
Poindexter
Poston, Charles D.
Powell, John Wesley
Prescott
Public Lands Commission
Puerto de San Augustine
Pumpelly, Raphael

Quaife, Milo Milton
Quartz Mountain Mining District
Quiros, Jesus
Quitovac

Ramlas, Rafael
Rastel, Jacob
Ray Mine
Raymond, Rossiter W.
Real de Dolores
Reddick
Reed, James
Rich Hill
Richardson Mining District
Richmond, William A.
Rio Bravo del Norte (Rio Grande)
Rio Conejos
Roberts, John
Roberts, William
Robertson, James O.
Robinson, Douglas
Robinson, Louie
Rodríguez, Agustín
Rose, Dan
Rowell, C. W. C.
Rusk, Theodore Green
Ryan, Cornelius
Ryther, Frank D.

Sacation (Sacaton) Station
Sacaton
Salero Mine
Salgado, Juan Jose
San Francisco Mining District
San Francisco Morning Call
San Francisco Vein
San Gabriel Vein
San Juan Company House
San Juan
San Pedro River
San Xavier Mission
Sanford, J. M.
Santa Adila Claim
Santa Catalina Mountains
Santa Cruz
Santa Cruz River
Santa Fe
Santa Maria River
Santa Rita Mines
Santa Rita Mountains
Sargent, Aaron A.
Schuchard, Charles
Seattle Mining District
Sherman, Caleb
Sherman, John
Sherman, William Tecumseh III
Shoup, Soloman

Names Associated with Creation of Mining Districts

This portion of appendix C was prepared as an aid to those seeking to identify those individuals whose names appear in the formation documents of the respective mining districts. This listing follows the timeline of appendix B so that the travel of miners between mining districts is more apparent. A note of caution—the deciphering of the spelling of names in handwritten records is extremely difficult, so reliance on the spelling of names below may be problematic. I thank my friend Jay Spehar for his assistance in reviewing my initial interpretations.

Castle Dome Mining District
Herman Ehrenberg
Jacob Snively
Geo. W. Leigh
Oliver Lindsay
Thomas Bidwell (maybe Thomas J.)
Abraham R. Hoagland
William P. Miller
W. L. Hopkins
J. D. Rittenhouse

Colorado Mining District
C. W. Bush
H. B. Coit (maybe B. H.)
Mr. Lewis
Mr. Miller
Wm. Calez

Pioneer Mining District
S. Shoup
Jas.V. Wheelhouse
Jos. R. Walker
Jos. R. Walker Jr.
John Dixon
Jacob Linn
Jacob Miller
Jack Swilling
Frank Finney
S. C. Miller
George Blosser

A. C. Benedict
T. J. Johnson
B. Ellis
A. B. French
Chas. Taylor
H. B. Cummings
Wm. Williams
F. G. Gillahan
Jackson McCrackin
Rodney McKinnon
Felis Cholet
M. Lewis
Jas. Chase
George Coulter
Mr. Murray
Dobbins
Mr. Green
J. M. Sanford
McKinnie
Hulton
W. L. Griffin (maybe N. L.)
H. L. Hubbard
J. Marsh
Crane
Ingalls
R. Kram
Lockhardt
S. Ruder
Van Smith
F. G. Christie
A. W. Adams

Weaver Mining District, La Paz
C. B. Green
James S. Trimble
J. P. Newcomb
Jas Reed
J. B. Chevalier
W. B. Marshall
H. M. Oliver
Louis Robertson
Mr. Freman
Mr. Gird
W Hanford
W. Thompson
A. B. Smith
Geo. W. Leihy
G. C. Welch
C. G. Mellon
G. C. Cross

Weaver Mining District, Yavapai
Arthur M. Henry

Yavapai Mining District
A.B. Smith
Geo. W. Leihy
G. C. Welch
C. G. Mellon
G. C. Cross

Walker (Lynx Creek) Quartz Mining District
T. J. Johnson
Captain Bogart
J. V. Wheelhouse
V. C. Smith (Van?)
Solomon Shoup
Cal Dobbins
Major McKinney
Sandford
A. Thom
McCrackin
J. Brooks
J. E. McCaffre (maybe McCaffrey)
J. Rees

Hassayampa Mining District
Robert W. Groom
Van C. Smith
Robert McCoy

Quartz Mountain Mining District
John West
C. M. Dorman

Geo. Lunt
A. O. Noyes
H. Brooks
James G. Barney

Cerro Colorado Mining District
Wm Sim
John A. Mahon (maybe Jr.?)

Walnut Grove Mining District
F. M. Larkin

Wickenburg Mining District
Jas. Moore
Henry Wickenburg

Turkey Creek Mining District
Charles Taylor
W. C. Collier
T. J. Arnold
Henry Clifton

Bradshaw Mining District
Mr. Moore
George M. Willing
Max Solomon

Woolsey Mining District
John Roberts
James O. Robertson

San Francisco Mining District
B. Silliman
D. T. McCall
Wm France
Mr Russell
A. E. Davis
C. W. C. Roswell
M. Welly
Jas H. Nicholson
Morse
Vales
Thompson
Reddick
Col Aikens
E. E. Clough
Wm Trance
R. A. Rise
L. F. McCall
Wm France

Agua Frio (Agua Fria) Mining District
Milton Hadley
N. L. Griffin
Alfred Zimmerman
N. P. Pierce
W. R. Basham (maybe Bassham)
W. F. Banning
G. W. Smith
Jesus Munguis (maybe Munguia)
Joseph Lennon
G. Johnson
Simon Roblez (maybe Robles)
George Clinton
Jesus Otaro

Big Bug Mining District
Jackson McCracken
J. M Boggs (maybe V. J. M.)
John H. Marion

Wauba Yuma Mining District
Robert A. Rose
Daniel Smith
E. M. McCarthy (maybe M. E.)
Sam Knodle
N. H. Vosler
F. M. Jackson (maybe M. F.)
D. T. McCall (maybe D. B.)
J. H. Bateman
D. O'Leary
Jos Coulter
James P. Bull

Beaver Mining District
Robert J. Osborn

Pine Grove Mining District
Joe Melchoir

Martinez Mining District
Willard Rice
Jas. A. Farnsworth
George A. Kimble

Bradshaw Mountain Mining District
McMorris
Geo Monroe
Samuel Boll

Wallapai (Hualapai) Mining District
James Fleming
Samuel Todd
H. E. Davis
Geo Okay
Berry H. Spear
Wm Fee
R. C. Todd (maybe R. S. or R. E.)
James P. Bull
Geo Fisher
M. E. Doolittle
Dan H. Smith
Henry Hardy
Dunkel (maybe Ben F. or Thomas J.)
Hoffner
S. M. Atchison (maybe S. N.)
Mr Toll
J. Slarer
A. C. Haskell (maybe Alonzo)
H. F. Baker

Tiger Mining District
J. Reese
E. A. Gordon

Sedgwick Mining District
Pierce Dorgan
C. G. Terry (maybe Clara)
Mr. Cox
Mr. Hall
Mr. Kennedy
Pierce Dorgan
C. G. Terry
Samuel Hill

Fort Rock Mining District
E. C. Bradshaw
Robert Kelly
Wm. Burch (maybe W. M.)

Copper Mountain Mining District
James Ballard
John W. Reed
W. L. Rynerson
Joseph S. Yankie
I. H. Itone
T. Stevens
A. Sibel
Capt M. Joy (Miles)
J. Colwell
Owen Roberts
E. M. Pearce
J. W. Reed
Eugene S. Goulding
R. B. Metcalfe (Robert Boyer)
H. M. D. Hewett
Edwin Polk
Wm Bonnell
E. B. French
Samual J. Freudenthal
Wm F. Clarke
Wm Osborn
Sig Heisl

Cedar Valley Mining District
Sam. A Pearson
Andrew B. Peterson

Jas P. Bull
Caldwell Wright
N W. Clark
Wm Collins
John Long
W.H. Gellis, Sr.
S. Devine
G. Darden
Peter Olinger
Juan Espanosa
Santiago Altimara
Juan Feloya
Rich Holt
Louis Hoebel
J. A. DeLund
John Faley
George Myers

Papago Mining District
A. Brichta
Samuel Hughes
H. S. Stevens
William J. Osborn
Miguel Alvarez
A. Lizard
Thos Hughes
Jose Fontez
R. N. Leatherwood
L. C. Hughes
H. S. Stevens
Tully Ochoa & Co.

Owen Mining District
A. F. Owen
J. McCrackin
R. Gird (maybe Richard)

Truman Mining District
Wm. G. Boyle
J. C. Truman
John E. Magee
H. C. Hodge
A. C. Benedict
John Mansfield
Cornelius Ryan
Henry Mims
Joseph Kings
S. W. Carpenter

Smith Mining District
D. B. Rea
Jerome Sawyer
W. Hanson
William Gardner
Alva Smith
Jack Long
P. J. Hand
W. Elliott
H. G. Morse
R. T. Stibell
J. W. Coughlan
R. M. Choate (maybe R. H.)
T. McArthur (maybe Wm T.)
M. Stibell
D. B. Rea
Jerome Sawyer
R. H. Choate
H. G. Morse

Pioneer Mining District
Marshall Barnes (maybe M. C.)
Harry Jones
C. O. Brown
R. D. Dicky
Wm. G. Boyle
Richard Powers
J. A. Devine (maybe J. E.)
Harry Jones

Peck Mining District
Geo. T. Hogle
Henry A. Bigelow
George Opdyke

Globe Mining District
A. R. Hammond (maybe Andrew R.)
R. B. Metcalf (maybe Robert B.)
R. H. Choate
C. M. Shannon
I. Winters
M. Morris
John Harvey
N. L. Griffin
F. Tarbell
M. Ensena
Geo. Scott
Edwin Polk
C. E. Buck (maybe Charles E.)
Omer Whitlock
G. C. Noland
L. E. Jones
F. J. Morris
G. A. Swasey

Verdi [Verde] Mining District
S. V. Kell
J. D. Boyd
Edw O. Dougherty

Tyndall Mining District
Wm G. Boyle
John E. Magee
George Cooler
Thomas Davis
Sydney Carpenter
John Burns
Cornelius Ryan

Pima Mining District
Fred Drew
A. Brichta (maybe A.C.)
John B. Best
James M. Leggett
J. R. Baldwin
Gwen Rusk
Wm Manscoe
H. S. Lymons
A. Bricta
S. W. Carpenter

Arivaca Mining District
John McCafferty
Capt. E. P. Voisard
Ed Hudson
Wm. G. Poindexter (maybe S. or A.)
Jay Holden
J. A. Apperson

Aztec Mining District
Thomas Davis
Col. B. J. Hinton
John Mansfield
Wm. G. Boyle
Arthur S. Lovelock
John E. Magee
Richard J. Hinton
John E. Magee
J. W. Carpenter
J. A. Apperson

Tomb Stone Mining District
A. E. Scheiffelin
Richard Gird
E. L. Scheiffelin
Alies Reyes
Theo E. Walker
S. W. Carpenter
Fred Drew

Helvetia Mining District
B. Hilfe (?)
Fred Hilfe (?)
Albert Tibelel (?)
William Mallen
Thomas Billew (?)
Thomas Redlick (?)
S. W. Carpenter
HA. Lemeoo

Harshaw Mining District
H. H. Holt
J. W. Davis
J. C. Otis
D. T. Harshaw (maybe David T.)
S. W. Carpenter
C. N. Labance
Daniel P. Tomey
Murdock McKay
Harry Clay
C. T. Dunavier
James Tea
Michael Fagan

Tucson Mining District
Chas. Bell
A. C. Benedict
J. Magee
E. A. Chambelain
W. L. Campbell
W. L. Wakfield
Frank Sullivan
J. W. Everts
Joe Plummer
D. P. Lowell
J. E. Tallmadge
S. W. Carpenter

Old Hat Mining District
Louis DuPuy
W. G. Guild (maybe W. E.)
Wm Haueck
W. C. Deen (maybe W. A.)
Issac Kaurin
John Foutes
W. H. Timmens
S. W. Carpenter

California Mining District
C. A. Milner
David P Carr

Catarina (Catalina) Mining District
M. A. Baldwin
R. F. Straine
Colonel Bixby
Colonel Scott
Prof. J. B. Caldwell

Cochise Mining District
George W. Seay
P. Rusk
J. Taylor
J. Wool
M. Volmar
J. D. Fink
A. J. Mitchell
P. Rusk
J. Wool
S. W. Carpenter
W. A. McDermott

Silver Mining District
Warren Hammond
C. F. Birnham
Walter Millar
George D. Lyng

Plomosa Mining District
Geo. A. Ellsworth
C. L. Bancroft (maybe Clinton)
Peter Smith
J. W. Johnson

Greenlee Gold Mountain Mining District
Chapman
Mr. Bentz
Mike Fracy
J. H. Dorsey
I. D. Pinkard
John Brick
Richard J. Bailey
Chas Kimmear

Red Rock Mining District
J. M. Williams
O. J. Reed (maybe Otis J.)
M. C. James
Thomas J. Ward
Thos Cusack
Chas Van Valkenberg
John L. Lount
David B. Rae (also Rea)
Alex N. Childes
L. B. Dunne
Mather Crocks
Benett Roth
Henry Grinell (maybe Grinnell)
S. W. Carpenter
M. S. Makeky
A. K. Brewer
John Callore
T. B. Kerr
James Fine
M. A. Kerr
James Peterson
T. Beltran
Henry McGrandy
Trinidad Farriori
Santiago Beltran
W. D. Smith
Richard Thomas
Nick Simons
Jos Trewin
S. W. Carpenter, Recorder
H. Hatch, Deputy

Hartford Mining District
S. P. Overton (maybe Solomon P.)
C. W. Russell
S. P. Overton
S. W. Carpenter
Leo A. Reed

Yellow Stone Mining District
C. Noll
C. W. Russell
John W. Watson
Frank Parker
Geo. Knudson
S.P. Overton (maybe Solomon P.)
James Lippett

Golden Basin Mining District
S. W. Carpenter
H. Hatch

Pajarita Mining District
J. M. McArthur
J. B. Lowrey
W. E. Jervosse
J. A. Rockfellow

Palmetto Mining District
A. W. Osten
G. W. Atkenson
James Spudy
S. W. Carpenter

Minnesota Mining District
C. C. Jones
N. S. Lewis
S. F. Walker
C. C. Jones
John K. MacKenzie

Greaterville Mining District
E. N. Fish

Seattle Mining District
J. R. Rice
C. H. Snyder
Jones
Powers
Walker
Scott
Young
Johnson

J. S. Lavorino
Don Wendt
Jesus Quiros

Weaver Mining District, Mohave
B. N. Coleman
John H Weaver
John K. MacKenzie

Wrightson Mining District
L. W. Wakefield
S. L. Parke
E. R. Devoe
C. Kramer
H. Brodt
C. Coonte

Lost Basin Mining District
H. S. Devine
James Hodgon

Copper Mining District
Jns L. Gregg (maybe John L.)
John J. Hill
R. F. Kirkland
E. B. Kirkland (maybe Elliot B.)

Little Cottonwood Mining District
J. W. Grant (maybe W. J. or S. W.)
Wm Krempe
A. H. Hanson
J. B. Adams
Danl Garnes (maybe Daniel)
Dan'l B. Leary
Jas Noonan
Geo Wald
Henry McAlister
H. Stout
Ike Bould
W. W. Alden
J.W. Michael
Geo W. Brooks
Wm Chaffin
John P. MacKenzie

Richardson Mining District
Joseph A. Ashlock
Frank Duffee
Geo. Waltz
Dan'l Remington
James Fulton
Sam'l L. Bernie
Harry Moore
Henry P. Bennett
H. Stout
Jms. Basshart
Fred Bauer
J. E. Stevens
John K. MacKenzie

Silver Hill Mining District
Joseph Robinson
Wm Clark
A. J. Bowman
Y. B. Contins
Wm. J. Clark
J. T. Bowman
T. S. Moore
Jas Larson
M. M. Clarke
Jas. Shibell

Indian Secret Mining District
Louis Seabright
Caldwell Wright
Henry Schaefer
E. J. Godman

Superstition Mining District
W. A. Kimball
Lawrence P. Kimball
J. H. Possenny
W. E. Possenny
J. H. Ramsey
J. F. Lewis

Jas Hawks
Wm Boliths
E. A. Robinson
J. R. Morse
Chas Edwards
Samuel Grayson
M. L. Spear
Jas Coswell (maybe Thomas J.)
J. G. Peterson
Chas. L Hall

Warm Spring Mining District

J. F. Burnes
C. F. Cox
L. S. Duham
Joseph Millett, Jr.
J. M. Davis
J. P. Nielson (maybe Joseph P.)
Byron G. Millett (maybe Byron A.)
Azro DeMill
Thomas P. Jensen
C. F. Cox
C. A. Bush

Reed Mining District

W. T. Donovan
P. P. Sullivan
Frank Foran
F. W. Jordan
Chas. W. Reed
John Caley
F. H. Hereford (maybe Frank H.)
Chas. A. Shibell

Greenfield Mining District

N. P. Greenfield (maybe H. P.)
W. B. Heslin
W. Warford (maybe Harold W.)
Chas. Shebell (maybe Chas A.)

Purtyman Mining District

J. F. Demick
John J. Thompson
Joseph L. Lay
J. E. Purtyman
Dan J. Cronin

Notes

Preface and Acknowledgments

1. Lacy, "Least Welcome Profession"; Lacy, "Going with the Current," chap. 10.

Chapter 1: Prologue—Legends and First Attempts at a Mining Law for the New World

1. The fruits of conquest were not easily enjoyed, as the jealousy and greed of competitors were constant considerations. See, generally, Pagden, *Hernán Cortés*. The spoils were not that great either. According to Cortés's accounts, 30,000 pesos were returned to Spain—representing the entire fruits of the initial conquest of Mexico. Subsequent captures netted an additional 130,000 pesos. Pagden, *Hernán Cortés*, 28, 265, and 492n79. At this time a peso de oro was 1/50 of a marc of 22 carat gold or 4.18 grams of gold (31.1 grams in a troy ounce), giving a total of approximately 21,505 troy ounces of gold—not a bad haul, but hardly worth the hardships that Cortés endured. In the case of Pizarro, his tactic of holding the Inca king Atahualpa for ransom successfully collected a room full of gold that was supposed to measure as high as the king could reach and eventually amounted to 1,326,531 pesos de oro or 7.43 tons of gold. See Prescott, *History of the Conquest*, 421, 428, and 453.

2. The Commission of Pánfilo de Narváez can be found in *Colección de documentos inéditos relatives al descubrimiento, conquista y organización de las antiquas posesiones españolas de América y Oceanía, sacados de los archivos del Reino y muy especialmente de las Indias*, series 1, 10:40 and 26:67.

3. See Clissold, *Seven Cities of Cíbola*, 48.

4. The literal translation is "head of a cow," and lest we think that Señor Núñez was some sort of eccentric, this was a title bestowed by the king of Spain after the battle of Las Navas de Tolosa in 1212 when a maternal ancestor marked a trail with a cow's skull, enabling the Spaniards to outflank and defeat the Moors. See Clissold, *Seven Cities of Cíbola*, 39.

5. Estevanico's race seems to be the subject of some debate. Marcos de Niza refers to him as a "negro," while the diarist with the Coronado expedition calls him a *moreno*, or brown man. Cabeza de Vaca, who should have known, says simply that he was an Arab. Wagoner, *Early Arizona*, 47.

6. Marco de Niza's report was an extreme exaggeration, but in fairness, never mentioned gold. Flint, *No Settlement, No Conquest*, 36–38.

7. Flint, *No Settlement, No Conquest*, 39.

8. Flint, *No Settlement, No Conquest*, 45.

9. See generally Wagoner, *Early Arizona*, 48–52.

10. Flint, *No Settlement, No Conquest*, 48.

11. Clissold, *Seven Cities of Cíbola*, 166; Wagoner, *Early Arizona*, 50–63.

12. Flint, *No Settlement, No Conquest*, 133.

13. See Aiton, "Ordenancas," 79–80; and Aiton, *First American Mining Code*, 108, citing Archivo General de Indias, 58-5-9, carta del Audiencia, Mexico, D.F., December 16, 1577. These provisions are described in detail in Lacy, "Going with the Current," § 10.03, 10–23 to 10–33.

14. Aiton, "Ordenancas."

15. See, generally, Wagoner, *Early Arizona*, 64–65.

16. On the discovery of silver mine prospects, see reports of Pedro de Bustamante, Hernando Gallegos, and Phelipe de Escalante as translated in Bolton, *Spanish Exploration*, 142 and 154.

17. Bartlett, "Notes upon the Routes."

18. Bolton, *Spanish Exploration*, 187.

19. Bolton, *Spanish Exploration*, 193 and 195.

20. Bancroft, *History of Arizona*, 110–27.

21. Most historians have suggested that the mines were either in the eastern slopes of the Aquarius Mountains or in the Hualapais. See "Account of the Discovery of the Mines, 1599," in Bolton, *Spanish Exploration*, 245n2; Peplow, *History of Arizona*, 1:192. The clues to their location begin with the account made of the trip in "Relaciones que envió Don Juan de Oñate de algunas jornadas," 11–20 as translated in Bolton, *Spanish Exploration*, 243–45. The account indicated that the party, after traveling along mountain ranges and valleys of northern Arizona, camped on the bank of a river. From this point they traveled four leagues through fertile land when they came to another river, wider than the first, that flowed almost from the north. They then traveled two leagues to another, much larger river that also flowed from the north. This river was crossed, and the party arrived at the slopes of some hills. It was reported that the banks of the rivers were seen with deep ravines, the finest of pastures, and extensive plains. They camped on the slope of the hills at a spring of water that was "almost hot." The Indians afterward drew a pattern of the three rivers "and two other which joined them further on, all united and passed through a gorge which they pointed out to them, and that beyond the gorge the river was extremely wide and copious, and that on the banks on both sides there were immense settlements of people who planted very large fields of maize, beans, and gourds in a very level country of good climate" before finally flowing into the sea. In a study of Arizona's thermal springs published in 1981, there are two hot springs that could have been the site mentioned in the account. The first is along Burro Creek near the confluence of Kaiser Spring Canyon (in the Southeast Quarter of the Northeast Quarter, Section 10, Township 14 North, Range 12 West, G&SRM [Gila and Salt River Meridian]) and the second slightly further to the north along the Big Sandy River near Wickiup (in the Southeast Quarter of the Northeast Quarter of the Southwest Quarter, Section 25, Township 16 North, Range 13 West, G&SRM). See Witcher, "Thermal Springs of

Arizona," 2–3. The latter of the two positions seems the more promising. This would make the first river crossed Spencer Creek (or San Francisco Creek) and the "much larger" river being the Big Sandy. The description of the valley would thus seem logical, the gorge the Indians pointed out would thus be at the western end of the Arrastra Mountains, the two rivers that later joined being Burro Creek and the Santa Maria River, and the wide plains being the fertile valley along the Bill Williams Fork. Both of these areas saw early mining activity. A narrative of the find of "rich silver ore" can be found in Bartlett, "Notes upon the Routes," 34–35.

22. Kessell, *Remote Beyond Compare*, 172.

23. A Castilian legua was equal to 5,000 varas. The Arizona map of mining districts indicates that such a position would have been in the Baboquivari Mining District.

24. Polzer, "Legends of Lost Missions," 173–75; Officer, "Mining in Hispanic Arizona." Officer points out that the Jesuit records do not indicate the existence of a priest named Escalante, but that a Franciscan, Silvestre Vélez Escalante, was stationed in New Mexico in the 1770s. Officer, Mining in Hispanic Arizona," 11.

25. Mining camps were called reals, and the Real de Arizona and the Arizona Mountains are shown on the maps of the day.

26. See Polzer, "Legends of Lost Missions," 177–80, based on an analysis of documents from Archivo General de Indias, Seville (AGI, 67-4-45, or Guadalajara 185). Other sources include a report by H. G. Ward, the charge d'affaires in Mexico from Great Britain in 1825–27, as published in Wilson, *Mexico and Its Religion*; Pumpelly, "Mineralogical Sketch," 2:138; Hamilton, *Resources of Arizona*, 144–45; and Wagoner, *Early Arizona*, 102–3.

27. An interesting parallel to the creating of the *planchas* exists in the processing of ore from the copper deposit at Santa Rita del Cobre, where molten copper was poured into clay-lined pits, the slag skimmed, and the rough ingots collected. This would suggest that it was a known practice and supports an argument that the *planchas* were the result of a smelting process. See Huggard and Humble, *Santa Rita del Cobre*, 16.

28. Bancroft, *History of Arizona*, 400–401n43. A "marc" was 230.0465 grams of silver, and an "arroba" was 12.5 kilograms.

29. See Browne, *Tour Through Arizona 1864*, 271 (who reported twenty-five mines), and Pumpelly, *My Reminiscences*, 1:133 (who stated that there were twenty-five mineralized veins).

30. Pima County Old Records, Tucson, Arizona, Book "A":1.

31. Bancroft, *History of Arizona*, 386.

32. The interviews of local inhabitants after the Pima revolt provide the best indication of actual mining operations. These interviews, together with other contemporary statement suggest that there were five mines in the area of the Santa Rita Mountains within the "Mineral Segregation" of the Baca Float No. 3 Private Land Grant, three mines in the Arivaca area, one mine near Guevavi Mission on the eastern slope of Mt. Benedict (northeast of Nogales), and one mine in the Southern end of the Huachuca Mountains. See Officer, "Mining in Hispanic Arizona," 4, 5, and 13n25.

33. These included a claim of José Lorenzo de Ribera on January 29, 1756 ("in the foothills on the other side of the Rio Grande"); Francisco García Carabajal on January 18, 1759 (near the Puerto de los Organos on the other side of the river"); and José de la Sierra on January 18, 1759. The claim of Francisco García is notable in that it lists twenty-one individuals who apparently had an interest in the claim. See Hendricks, "Spanish Colonial Mining," 1143–46.

34. It was during this time that the principal mines were developed except for the very limited scale near Tubac; see Bancroft, *History of Arizona*, 400. The governors of the Mexican provinces were replaced in 1786 by governor intendants under the authority of the viceroy, the commandant general, or the *audiencia*. The intendance of Arizpe included Sonora and Sinaloa. In judicial matters, Sonora, like other parts of northern Mexico, came under the *audiencia* of Guadalajara. Bancroft, *t*, 1:670–72, 674–75, 676–79; Beers, *Spanish and Mexican Records*, 312.

35. The copy was transmitted on January 24, 1788, by Jacobo Ugarte y Loyola at Arizpe in Sonora to the governor of New Mexico, Spanish Archives New Mexico II, roll 11, frame 36.

36. Bancroft, *History of Arizona*, 399–401. Bancroft reports that in 1828 some poor people from a rancho near Calabazas were working a gold mine. This may be a reference to Guevavi.

37. Officer, "Mining in Hispanic Arizona," 7.

38. Tom Childs Sr. reported evidence of Spanish operations at Ajo in 1836. Rose, *Ancient Mines of Ajo*, 14–18. It has been suggested, however, that evidence at the Arizona Historical Society indicates that Childs did not arrive in Arizona until 1857, when the mine was already in operation. Officer, "Mining in Hispanic Arizona:," 7. Whether or not Child's report is accurate, however, the Ajo copper deposit was clearly known as a mine when Peter Brady, a member of the boundary survey party, examined the site in 1854. Gray, *Survey of a Route*, 215–16.

39. See generally, Huggard and Humble, *Santa Rita del Cobre*.

40. See Hendricks, "Spanish Colonial Mining," 159. Twitchell, *Facts of New Mexican History*, 1:475; Jones, *New Mexico Mines and Minerals*.

41. Beers, *Spanish and Mexican Records*, 313–14.

42. See Quaife, *Christopher Carson*, 9–21, for a description of how this authority was circumvented.

43. Bowden, "Private Land Claims," 491–98.

44. An Act to Confirm Certain Private Land Claims in the Territory of New Mexico, chap. 66, 12 Stat. 887 (1861). The status of the surrounding lands was the subject to continuing litigation but a tract of 69,458.33 acres less 259 acres covered by the Cañon del Agua Grant was eventually patented on May 20, 1876. See Bowden, "Private Land Claims," 495–98.

45. Jones, *New Mexico Mines and Minerals*, 23.

46. Spanish Archives New Mexico I, # 1143.

47. Spanish Archives New Mexico I, # 1143.

48. Cozzens, *Marvelous County*, 175.

Chapter 2: The Mining Law in the Wilderness of Arizona

1. The "takeover" was portrayed as peaceful, but subsequent uprisings in Taos and Santa Fe during the winter of 1846–47 suggest otherwise. See Gómez, *Manifest Destinies*, 28–31.

2. See the "Disturnell Map" of the Republic of Mexico, which was used as the basis of the Gadsden Treaty. This problem is illustrated in Walker and Bufkin, *Historical Atlas of Arizona*, 18–19.

3. Kearny may have overstepped his authority in undertaking this effort, particularly in declaring New Mexico as "a territory of the United States," but writers suggest that he was probably following orders from the president; see Gómez, *Manifest Destinies*, 23. The citizens of "Arizona" almost immediately began lobbying for a separate territorial status. In 1856, two memorials were sent to Congress to that effect. The first, signed in Mesilla on January 23, 1855, by W. Claude Jones (who later become the speaker of the First Arizona Territorial Legislature) and fifty-six others, and the second, signed in Tucson later that year, suggested a division to the south along 34 degrees 20 minutes north latitude. Larson, *New Mexico's Quest for Statehood*, 88. A "formal" convention for this separate territory was proclaimed at a meeting in Tucson on April 2–5, 1860, by thirty-one citizens disgruntled with the New Mexican territorial government. This territory included the area of Arizona and New Mexico south of 33 degrees 40 minutes north latitude. This same group was also partially responsible for the establishment of a short-lived Confederate Territory of Arizona, which was proclaimed by Jefferson Davis on February 14, 1862, ironically exactly fifty years prior to Arizona's formal statehood status. The Confederate territory included the area of New Mexico Territory south of 34 degrees north latitude. These "wildcat" territories had named Mesilla and Tucson, respectively, as their territorial capitols. See Wagoner, *Early Arizona*, 443–56; and Walker and Bufkin, *Historical Atlas of Arizona*, 28–29.

4. Dawson, *Doniphan's Epic March*, 2. Doniphan was assisted in this effort by Willard Hall, an attorney from Missouri, and David Waldo, a Yale graduate, *Doniphan's Epic March*, 85.

5. See generally, Mumey, *Notes to Kearny Code*. According to one scholar, the code "established American institutions and introduced common-law concepts." Bakken, *Development of Law*, 23. Another has described the code as "compounded of Mexican, Texan, and Coahuila statutes, the Livingston Code of Louisiana, and the organic law of Missouri," and as creating "a governmental structure for New Mexico similar to that envisaged by the Northwest Ordinance of 1787." Lamar, *Far Southwest*, 63.

6. The ores were transported to Mexico City by pack mules and wagons. See Huggard and Humble, *Santa Rita del Cobre*, 19; see also Twitchell, *Facts of New Mexican History*, 1:475 and 2:37; and Christiansen, *Story of Mining*, 19.

7. Kearny Code, 83, Laws, section 1. The effort drew considerable criticism as exceeding Kearny's authority, particularly in its conciliatory approach to existing rights and the application of the Constitution of the United States. Dawson, *Doniphan's Epic March*, 82–83.

8. Kearny Code, 95.

9. Kearny Code, 114. Water Courses, Stock Marks, &c., section 1.

10. His secretary of state, George W. Halleck, a lawyer and later general-in-chief of the Union Army during the Civil War, did not sign the proclamation as he believed that the right to public land needed action of the United States. See, generally, Yale, *Legal Titles to Mining Claims*, 17, and Lacy, "Historical Overview of the Mining Law."

11. See Lacy, "Going with the Current," chap. 10.

12. This reference is probably to the mountain range. Gray, *Survey of a Route*, 78.

13. Gray, "Reminiscences of Peter R. Brady," in *Survey of a Route*, 215.

14. Pumpelly, *My Reminiscences*, 1:248.

15. Report of New Mexico Convention of Delegates 19 (Feb. 25, 1850).

16. New Mexico Territorial Laws (1851), 13.

17. New Mexico Territorial Laws (1851) Joint resolution, 1st Sess., 227.

18. This uncertainty is not to suggest that they were lost; the boundary itself was the subject of dispute between the United States and Mexico. See Wagoner, *Early Arizona*, 281–97, and particularly the map on 280.

19. The Gadsden Treaty between the United States and Mexico, Dec. 30, 1853, 10 Stat. 1031.

20. Blake, *Report of the Governor*, 48–49, quoting Peter Brady.

21. Doña Ana County Records Book B, Las Cruces, New Mexico.

22. *London Mining Journal*, November 7, 1857, 771.

23. *London Mining Journal*, November 21, 1857, 807.

24. Mineral Survey 2679, for the Midnight Show, Mayflower, Lone Star, Hermosa, Hot Head, Ajo Queen, and Wedge claims. These claims were subsequently transferred to the New Cornelia Copper Company.

25. Doña Ana County Records Book B, page 41.

26. Doña Ana County Records Book B, page 102.

27. Doña Ana County Records Book B, pages 103, 107.

28. Doña Ana County Records Book B, page 130.

29. Doña Ana County Records Book B, page 221.

30. Doña Ana County Records Book B, page 319.

31. Doña Ana County Records Book B, page 361.

32. Doña Ana County Records Book B, page 442.

33. Walters was Gray's assistant surveyor that summer and fall. *Weekly Arizonian*, August 4, 1859, 2:2.

34. Doña Ana County Records Book B, pages 456, 458. In the location notice, it states that the Indians with Gray said that "no white man had been there [the Maricopa Mine] before." It is possible that the deposit has been "worked" by indigenous people. Gray, Walkers and Bryant were surveying the Pima-Maricopa (Gila River) Indian Reservation between September 5 and October 17, 1859, which shows them closely working with tribal members; see Gray, "Annual Report."

35. Doña Ana County Records Book B, page 459. Gray's hesitancy undoubtedly grew out of his experience as a government surveyor for the Mineral Land Agency

under the secretary of war in the Michigan copper country, where mineral deposits were leased under the federal leasing act of March 3, 1807. By 1845, President Polk, in this message to Congress, said that the system was "believed to be radically defective," and it was accordingly repealed. See Curtis Lindley, *Treatise on the American Law*, vol. 1, sections 33–36, pp. 67–69. Gray's letters and actions during the 1840s seem to reflect his knowledge of the uncertainty at the time and that the laws would change. See Gray, letters in Fedner, *Fort Wilkins 1844*.

36. The 1907 U.S. Land Monument #2395 placed at Gray's mine site is at 33 degrees 05 minutes north latitude and 111 degrees 06 minutes west longitude—the inaccuracy of the 1860 filling is the possible product of transcription or clerical error; The elevation of the U.S. Mineral Monument #2395 is roughly 3,300 feet, where Gray gave the elevation as 3,378 feet, an acceptable margin of error for the time. Gray's 1860 mine report is reprinted in Browne and Taylor's *Reports upon Mineral Resources*, 451; Browne also reprinted Gray mine reports by Brunckow and by Hopkins, and optimistically says "This lode, sometimes called Gray's mine, situated . . . four miles south of the Gila river, is considered one of the best copper deposits in southern Arizona."

37. Doña Ana County Records Book B, page 460. After the Civil War, several parties relocated the then abandoned mine until, in the 1890s, a Cornish miner, William Mill Williams, relocated the "old Gray mine" (Maricopa) as his Eureka No. 4 claim (Pinal County Recorder's Office, Book 15 of Mines, pp. 213–14, 197, 191–92). He undertook the first serious mining on the claim, and with his eastern backers Alba Kent and James Rothwell received a patent February 11, 1910. Because he had a peg leg, Williams' mine (Gray's Maricopa) was also called the Peg Leg Mine. See Peg Leg Mine file online on Arizona Geological Survey: https://mininginfo.azgs.arizona.edu. More recently, the mine site was owned by the Timios Prodromos Ecclesiastical Retreat Trust, located a mile or so off the Florence-Kelvin road at the end of Peg Leg Road.

38. Doña Ana County Records Book B, page 460. Doña Ana County Records Book B, page 599.

39. "Reminders of Day When Arizona Was a Confederate Territory," *Arizona Republican*, January 1, 1922.

40. The Spanish *leuga* varied, but in the Spanish Southwest surveys were about 2.63 miles.

41. Doña Ana County Records Book B, page 460.

42. Doña Ana County Records Book B, page 570.

43. Doña Ana County Records Book B, pages 563 and 613.

44. Mowry, *Arizona & Sonora*, 76. Wagoner, *Early Arizona*, 423, reports that Moore, Lord, and Ewell had originally purchased the mine from a Mexican herder in 1857.

45. Doña Ana County Records Book C, page 108, 185.

46. Walker, "Confederate Government in Doña Ana County," 253–302.

47. The acts of the New Mexico Territorial Legislature are included in separate volumes by year but are not individually numbered. The legislation referred to above

is found in the Acts of February 1, 1860 (section 1), January 8, 1862 (section 2, page 16), and January 28, 1863 (page 30). This indecision may have been influenced by the creation of the Confederate Territory of Arizona on February 14, 1862.

48. Poston reported that "Spanish and Mexican grants hung over the county like a cloud, and the settler could not be certain of a clear title." Poston, *Building a State*, 49. Similar sentiments can be found in a proposal for separate territorial status for Arizona, Mowry, *Memoir of the Proposed Territory*. Old Records, Book A, Pima County, 1–10, contains a registration of title of the Aribac land grant dated August 4, 1824. Page 11 is a deed from Ignacio Ortiz to Aribac "in demarkation of Tubac" to Samuel P. Heintzelman, U.S.A., president of Sonora Exploring and Mining Co., for two square leagues "with all the wood, water, grass, ores, mines and minerals on the same," for consideration of "25 shares and Dollars." The document contains a notation that it was filed in the General Land Office for the department of Pimeria Alta on July 2, 1833, at Arispe, Sonora.

49. The Pima County records indicate in Book A that James A. Lucas, clerk of Probate Court of Doña Ana County, appointed Poston as a deputy clerk on July 15, 1856.

50. In this context, Poston claimed that during the late 1850s in Tubac there was no law but the law of love. Poston, *Building a State*, 74. The matter of the illegal marriages was supposedly resolved by the local priest who remarried all participants in the civil ceremonies along with Poston's promise to cease and desist. Posten's functions as a justice of the peace may have been slightly exaggerated, as Poston's record book shows the issuance of only one marriage certificate, Pima County Old Records, Book No. A, page 203, although a number of pages are missing.

51. The mine was reported to have been worked 120 years earlier under the superintendence of the Jesuits, then living at the mission of Tumacacori. The settlers had apparently called it the salt-cellar mine, because the bishop supposedly required a source of salt for the mission. Cozzens, *Marvelous County*, 173–74. These efforts were generally considered at the time to have been worked earlier by Spanish explorers, and, although generally considered valuable, had been abandoned on account of Apache interference. Cozzens, *Marvelous County*, 175.

52. This determination was finally affirmed by the United States Supreme Court. Lane v. Watts, 234 U.S. 525 (1914). The Department of Interior was convinced from the beginning that the land was "mineral in character" and had sought to segregate land from the land grant selection and bitterly fought Watts's application in at least nine separate opinions. The Supreme Court's ruling, however, was that a finding made in 1866, regardless of its validity, was final.

53. Pima County Old Records, Book A, pages 195–196.

54. Pima County Old Records, Book No. A, page 197; Pima County Old Records, Book No. A, page 197, 199.

55. A "vara" varies in length but is generally thirty-two inches.

56. North, *Samuel Peter Heintzelman*, 120, 156.

57. North, *Samuel Peter Heintzelman*, 155.

58. See the description of the operations in and around the Santa Rita silver mines

in Pumpelly, *My Reminiscences*, 194–249. Life was indeed difficult at the Santa Rita Mine, and during an eighteen-month period in 1860–61, five managers of the mine were killed by either Apaches or Mexican bandits.

59. Raphael Pumpelly reports, as an employee at the Santa Rita Mine, that the lease had been transferred to Samuel Colt at this time. Colt had been one of the principals of the company, which suggests the prior arrangements had never been formalized. Pumpelly, *My Reminiscences*, 1:238.

60. Paul, *Mining Frontiers*.

61. Burggraaf, *Across New Mexico and Arizona*.

62. Etter, *To California*, 36.

63. The actual "rush" may have been sparked when Paulino (Pauline) Weaver showed gold dust to José María Redondo, a trader and miner at the Gila City diggings. Redondo, Felipe Amvisca, Antonio Contreras, and Jesús Contreras may have been the true discovers. See the comments by Alexander McKey, a member of the territorial legislature from La Paz, reported to Browne and Taylor's *Reports upon Mineral Resources*. By February 1862, the word had spread, but the interest was primarily within the Hispanic community. During April and May, the fever had hit Los Angeles as samples of gold reached the city, and Juan Ferrar, a forty-eight-year-old Sonoran, had picked up a specimen reported to be the size of a "hen's egg." Then George Hooper, the trader at Fort Yuma, showed up in Los Angeles with $8,000 in gold dust. See Thompson, "Henry De Groot," 37:134–35.

64. April 5, 1869, letter in Langley, ed., *To Utah with the Dragoons*, 150.

65. The resolutions were to be published in the *Washington Union*, which was reprinted in the *Pittsburgh Post* on December 13, 1858.

66. By then the population was rushing off to a new gold strike near the headwaters of the Gila River at Pinos Altos north of the present location of Silver City. Gila City would be obliterated by a flood in January 1862, as described by a correspondent of the *Daily Alta California* January 27, 1862, and reprinted in the *Nevada Democrat*, Nevada City, February 18, 1862.

67. These expeditions were described in the press—the first in the *Weekly Arizonian* on June 9, 30, 1859, and the second in *San Joaquin Republican* on April 12, 1860. Snively wrote letters to friends and the press to announce the May 1860 discovery of gold at Pinos Altos: in the *San Francisco Weekly Herald*, June 14, 1860, Snively reported that, about eighteen miles from the Santa Rita del Cobre copper mines on Bear Creek, "we have found mines of gold and silver. The gulch in which we struck the gold is very rich, paying on average twenty cents per pan, some of the dirt paying as high as a dollar. The section of gold bearing country is extensive, and thousands of poor men can find employment."

68. See the comments by Alexander McKey, a member of the territorial legislature from La Paz, reported to Browne and Taylor, *Reports upon Mineral Resources*, 453–54.

69. See Thompson, "Henry De Groot, 1862," 37:134–35.

70. Wm. Bradshaw, report in *Los Angeles Star*, June 14, 1862. Another writer noted that Sonoran women were also miners, working alongside their male counterparts.

71. *Daily Alta California*, May 18, 1862.

72. La Paz District Records, Book 1 Claims and Deeds, page 1.

73. The reference to "dips, spurs and angles" was frequently employed in California and elsewhere in the mining frontier to describe what became known as the "extralateral right," the right to follow a vein on its downward course laterally underground beyond the surface sideline boundaries of the individual claim. The three locators of the rich Yellow Jacket claim at Gold Hill, Nevada, for example, claimed, "Twelve hundred (1200) feet of this Quartz Vain including all of its depths & Spurs." As Dan DeQuille commented, "what the locators meant by 'depths' in their notice was dips; no matter in what direction the 'vain' might dip, they desired to put on record their right to follow it." DeQuille, *Big Bonanza*, 35–36. While William Stewart, senator from Nevada, claimed credit for drafting the first California district regulations defining a vein or lode claim as consisting "of 200 feet in length along the vein or lode, with all its dips, spurs and angles," he was probably not the first to employ this terminology. Brown, *Reminiscences of Senator Stewart*, 127. While a mining lawyer on the Comstock, Stewart recognized the "difficulties of the contest and the extraordinary expenses which must be incurred" in the course of "extralateral rights" litigation over the claims which made up the Comstock Lode. He proposed a settlement by which the parties would have agreed "that the mining operations of all companies should be confined within the limits of the planes bounding their respective claims . . . a relinquishment of the cherished but litigious principle which allowed a locator to follow the dips of his ledge indefinitely, and a substitution of the often-decried Spanish or Mexican system of allotment." Lord, *Comstock Mining and Miners*, 144–45. Though Stewart's effort at compromise failed on the Comstock when one of the parties refused to accede to the proposal, he effected a similar compromise in Tombstone in 1881 with the owners of the Contention Mine. Brown, *Reminiscences of Senator Stewart*, 266–68 and Tombstone's *Weekly Epitaph* (October 15, 1881). While many miners coming to Arizona from other mining districts throughout the West may have understood the language of "dips, spurs and angles" as connoting, in Dan DeQuille's phrase, "their right to follow" a vein "no matter in what direction . . . [it] might dip," it should be noted that only one of the Arizona district regulations expressly granted this right. Most simply referred, without elaboration, to the "dips, spurs and angles" as part of the claim to be established under their tenets.

74. Castle Dome Mining District Regulations as published in King, *United States Mining Laws*, 247–51.

75. Castle Dome Mining District Regulations as published in King, *United States Mining Laws*, 247–51.

76. By the next year, the number of recorded claims in the district was reported to be 396 (inclusive of extensions) upon 150 separate ledges. See Blake, *Diary*, January 19, 1864, vol. 63.

77. "Record of No. 1 Quartz Mine," La Paz Records, Book 1 Claims and Deeds, page 1.

78. La Paz Records, Book 1 Claims and Deeds, page 82.

79. "Notice 30 Ehrenberg Lead," La Paz Records Book 1 Claims and Deeds, page 26.

80. "Colorado Vein (No. 65)," La Paz Records, Book 1 Claims and Deeds, page 92.

81. "The Eugenie Quicksilver Mine No. 83," La Paz Records Book 1 Claims and Deeds, page 136.

82. North, *Samuel Peter Heintzelman*, 151.

83. Colorado Mining District Regulations, Mohave County Historical Society Archives (Kingman, Arizona).

84. Colorado Mining District Regulations, section 5.

85. Colorado Mining District Regulations, sections 6 and 11.

86. Colorado Mining District Regulations, section 8.

87. Colorado Mining District Regulations, section 10.

88. Colorado Mining District Regulations, sections 12, 14, and 16.

89. The Freeman District was described as "commencing on Colorado R. opposite Needle Point running westerly 20 miles, thence southerly 30 miles, thence easterly to the Colo R., thence down the R to the place of beginning." From the June 12, 1863, regulations, William McKay, recorder. Blake, *Diary*, January 15, 1864, vol. 62.

90. Weaver was the son of a Tennessean father and a Cherokee mother and was apparently originally named "Powell." While in Taos, New Mexico, he began using the Spanish derivation "Paulino," which shows up as late as 1864, where a handwritten location notice (seemingly dictated to Territorial Secretary Richard McCormick, as Weaver usually signed his name with an "X") was signed as "Paulino"—although "Pauline" also appears in the notice. Also, it seems incongruous that a fellow occupying Weaver's position could get through life known as "Pauline."

91. Weaver Mining District Regulations (Yuma County), in King, *United States Mining Laws*, 252–53.

92. *San Francisco Evening Bulletin*, September 29, 1863.

93. King, *United States Mining Laws*, 251.

94. *Arizona Weekly Miner*, December 29, 1876, 4:1.

95. Melchiorre, *Gold Atlas of Rich Hill*, 14–16; and Crombie et al., *Rich Hill*, 39–41.

96. Pioneer Mining District Records, pages 2–4, Yavapai County Recorder's Office, Prescott, Arizona.

97. Hunt, *Kirby Benedict*, 83–84, 87. Albert Benedict's letter of May 21, 1863, stated that "I am the discoverer of the silver lodes and I believe I am entitled to the claim of the same. I am going to locate claims for some of my friends in Santa Fe. I send you a list of them. I wish to know if I can organize companies and locate claims for them and if the proceedings must be recorded in Santa Fe in order to secure legal rights." Hunt, *Kirby Benedict*, 6. Both Albert Benedict and Judge Benedict describe each other as being distant relatives. Hunt, *Kirby Benedict*, 84.

98. Hunt, *Kirby Benedict*, 87–88. The notice also included among its claimants the territorial officials and probably thereby set the precedent for Paulino Weaver's similar action the next year.

99. See generally Hinton, "Frontier Speculation." The initial miners' meeting, the selection of the offices of the district, and the drafting of the bylaws is described in Conner's account. The party initially wanted to appoint Joseph Walker as the "governor of the district," but he declined whereupon the position was offered to

Conner, who also declined on the ground that "he would not be there any longer than the first opportunity to get away presented itself," whereupon Thomas Johnson was appointed and accepted. Conner was also elected secretary but declined in favor of Wheelhouse, but Conner did finally consent to act as the scribe for the bylaws, and the initial entries in the recorder's book are in his handwriting. Conner, *Joseph Reddeford Walker*, 100.

100. King, *United States Mining Laws*, 253–54.

101. King, *United States Mining Laws*, 254.

102. King, *United States Mining Laws*, 254.

103. Walker Quartz Mining District Records, in King, *United States Mining Laws*, 258.

104. Recited as Cal Dobbins, A. Thom, and Jackson McCrackin.

105. King, *United States Mining Laws*, 255–56.

106. King, *United States Mining Laws*, 256. On Mexican placer miners, the employer of Mexican labor had to assume financial responsibility—pretty much a peonage system. Of course, in the Weaver District there was a threat of losing their claims and eviction if the employers (patrones) could not control their peons, to keep them in "good behavior." How long these rules were enforced is probably brief, suggested by editorials in 1865 by John Marion in the *Arizona Miner* praising the hardworking *placeros*.

107. Yavapai Mining District Regulations, Official Records, Yavapai County, Arizona; King, *United States Mining Laws*, 257.

108. Yavapai Mining District Regulations, article 6.

109. Yavapai Mining District Regulations, articles 8 and 9.

110. Hassayampa District Regulations; *Arizona Miner* 1:2, 3 (April 20, 1864).

111. Hassayampa District Regulations, article 3.

112. Quartz Mountain Mining District Records, 11–15, Yavapai County, Arizona; King, *United States Mining Laws*, 260–61.

113. Quartz Mountain Mining District Records, article 6.

114. Quartz Mountain Mining District Records, article 7.

115. For a description of the escape of Charles Poston and Raphael Pumpelly, see Wallace, *Pumpelly's Arizona*, chaps. 5 and 6, pp. 87–120.

116. Pima County Old Records, Book No. B, page 1, filed May 16, 1864.

117. Pima County Old Records, Book No. B, page 1, filed May 16, 1864.

118. Pima County Old Records, Book No. B, page 75. This may be the mine referred to as being near the Guevavi Mission. See Officer, "Mining in Hispanic Arizona," 5.

119. Pima County Old Records, Book No. B, pages 82–84. The claim on the "Chiricahua Lead" also included a claim to the water rights in a spring.

120. Pima County Old Records, Book No. B, pages 86–96.

121. Pima County Old Records, Book No. B, pages 78–80, 199–202. The "Cimarron Copper Mine" was recited to have been sold by Fritz Contzen to the locators, having been first located in conjunction with Julius Contzen in 1857, but had not been continuously worked owing to the "hostilities of the savage tribes." The peti-

tion to the probate judge requested that the claim be "registered and denounced . . . according to the Ordinances de Mineria in force in this Territory."

122. Cerro Colorado Mining District Regulations, *Arizona Miner*, 1:2, 3 (Aug. 24, 1864).

123. Cerro Colorado Mining District Regulations, article 5.

124. Cerro Colorado Mining District Regulations, article 9.

125. Such a provision was inconsistent with the general theme of mining law that measures the rights of the claimant from the time of posting on the ground (see Arizona Revised Statutes § 27–202), but is consistent with common law principles relating to the recording of conveyances that permits an assumption that the owner of record holds title until a conveyance or encumbrance is recorded. See Arizona Revised Statutes § 33–411.A ("No instrument affecting real property is valid against subsequent purchasers for valuable consideration without notice, unless recorded as provided by law in the office of the county recorder of the county in which the property is located").

126. During 1864, the Eureka Mining District was formed in Yuma County on January 2, 1864, and its regulations followed the pattern of the La Paz District. See King, *United States Mining Laws*, 261–62. In Mohave County, the San Francisco District was in existence, and minutes of a miners' meeting on October 17, 1864, have been found. The original records were not found, as is also the case with the Sacramento District also known to have been in existence in Mohave County. It is possible that this might be attributable to the fact that when the county seat was moved from Hardyville to Kingman, the records were "stolen" in the course of transportation.

127. Walnut Grove Quartz Mining District Regulations; King, *United States Mining Laws*, 263–64; Wickenburg Mining District Regulations, *Arizona Miner*, 4:2 (Jul. 20, 1864); King, *United States Mining Laws*, 263; Turkey Creek Mining District Regulations, *Arizona Miner*, 1:1, 2 (Aug. 24, 1864); King, *United States Mining Laws*, 266; Bradshaw Mining District Regulations, *Arizona Miner*, 1:1, 2 (Oct. 26, 1864); King, *United States Mining Laws*, 264–65; Woolsey Quartz Mining District Regulations, *Arizona Miner*, 4:1, 2 (Oct. 26, 1864).

128. Walnut Grove District Regulations, article 3.

129. Walnut Grove District Regulations, article 4.

130. Walnut Grove District Regulations, article 7.

131. Walnut Grove District Regulations, article 8.

132. Walnut Grove District Regulations, article 9.

133. See Wickenburg Mining District Regulations.

134. Wickenburg Mining District Regulations, article 3.

135. Wickenburg Mining District Regulations, article 6.

136. Wickenburg Mining District Regulations, article 7.

137. Wickenburg Mining District Regulations, article 11.

138. Turkey Creek District Regulations, article 6.

139. Turkey Creek District Regulations, article 11.

140. Turkey Creek District Regulations, article 12.

141. Turkey Creek District Regulations, article 13.

142. Turkey Creek District Regulations, article 16.

143. Bradshaw District Regulations, article 2, section 1.

144. Bradshaw District Regulations, article 2, section 2. The miners showed a remarkable appreciation of the fact that the territories were governed by the United States Congress and territorial laws would not supplant federal jurisdiction over the mines without some clear relinquishment of authority by the Congress.

145. Bradshaw District Regulations, article 2, section 4.

146. Bradshaw District Regulations, article 4, section 2. Similar provisions had been made by the Wisconsin legislature for parties erecting dams or other stream improvements aiding the Wisconsin timber industry. The explanation, in the words of one legal historian, was that "the community puts great store in releasing men's energies of mind and will through private property and contract. Normally, if an enterprise needed the use of another's land it must obtain the land in market, by bargain. When the legislature delegated to stream improvers the sovereign power to take land by eminent domain, it allowed the improver to insist not on a bargained but on a law-determined price. In doing so, the legislature attested how deeply it felt that law was properly used to foster investments which would multiply productive activity." Hurst, *Law and Economic Growth*, 172–73. By making these provisions for compensation, the district miners stood squarely in the Anglo-American tradition that emphasized both protection of property rights and the fostering of enterprises seen as especially beneficial to the community. As expressed by revolutionary spokesman Arthur Lee, "the right of property is the guardian of every other right." Reid, *Constitutional History*, 29. Taking his title from Lee's dictum, James W. Ely, in his *Guardian of Every Other Right*, likewise emphasizes that the right of property was enshrined in the Fifth Amendment to the United States Constitution, and that the right to just compensation for taking of property was seen by jurists of the early republic as "derived from natural law." At the same time, Ely notes a number of instances in which certain types or usages of property were deliberately fostered by colonial governments in preference to other types or usages. He cites the grant of the power of eminent domain to owners of gristmills and ironworks, establishments which "though privately owned, benefitted the community and were viewed as types of public service. Hence it was thought appropriate for such enterprises to possess the advantages of eminent domain." Ely, *Guardian of Every Other Right*, 24–25, 76.

147. Bradshaw District Regulations, article 5, section 1.

148. Bradshaw District Regulations, article 6.

149. Woolsey Mining District Regulations, article 8.

150. Woolsey Mining District Regulations, article 13.

151. Woolsey Mining District Regulations, article 14.

152. There do not appear to have been many disputes over ownership during the early years under the mining district regulations. This may have been due to the fact that most of the districts were groups of miners operating jointly and outsiders were simply not known. This was certainly true of Weaver's Peeples party, and the Walker party went to extremes to make sure their regulations didn't allow outside "interference." One dispute was recorded, however, in the Walker Mining District on March

6, 1864. The records do not disclose the nature of the dispute, but the interesting aspect is that apparently every miner in the district acted as the jury—the verdict going to the defendants by an 18 to 11 vote. See King, *United States Mining Laws*, 255.

153. *Arizona Weekly Miner*, 1 (Oct. 26, 1864).

154. United States v. Andres Castillero, 67 U.S. (2 Black) 17 (1863).

155. *Arizona Weekly Miner*, 1 (Oct. 26, 1864).

156. *Arizona Weekly Miner*, 1 (Oct. 26, 1864).

157. This claim had been located by Paulino Weaver on May 29, 1862, when he posted a notice on a copper lode which he called the "Aravaypa," located "15 miles west of Old Fort Breckenridge and 2½ miles southeast of the Cottonwood Spring on the Leach route and immediately on the wagon road from that spring to the Cañada del Oro." When Weaver formalized his discovery, he claimed three hundred feet for his discovery and three hundred feet for preemption "or as much as the law may allow." He also claimed three hundred feet each for the following individuals: James Carleton, Yager (leaving a space for the insertion of Yager's first name), John N. Goodwin, Richard McCormick, Henry W. Fleury, Benjamin Weaver, and Lauren Sparks. This location notice was formalized by a petition made in Prescott more than two years later to Richard McCormick as the secretary of the Territory of Arizona, and the record was made in McCormick's hand at the office of the secretary on August 22, 1864. See Archives, Office of the Arizona Secretary of State (Phoenix, Arizona). The naming of the territorial officials as claimants was obviously a politically inspired afterthought, as the territorial officials had not yet been appointed on the date of location. In fact, John A. Gurley was not appointed the territory's first governor until March 4, 1863, but died in August of that year. Goodwin, originally appointed as the chief justice of the Territorial Supreme Court, was then appointed governor. The naming of Carleton was more current, of course, as he was the commander of the California Column, which was providing protection from the Apaches at the time of Weaver's location. See Wagoner, *Early Arizona*, 473–74.

Chapter 3: Challenges to the District Rule

1. The party had to hurry to proclaim the new territory, as its enabling legislation required the establishment of the territorial government before the end of 1863. Another reason for the formality was that the territorial enabling act specified that the territorial officers could not be paid "until they have entered upon the duties of their respective offices within the said Territory." Instead of waiting until they got to Fort Whipple, their destination near present-day Prescott, the proclamation of statehood was therefore made at Navajo Spring, approximately six miles inside the new territory. Wagner, *Arizona Territory 1863–1912*, 32–33.

2. Act of February 24, 1863, 12 Stat. 662.

3. *Arizona Miner*, 1:2 (Jul. 20, 1864).

4. William T. Howell was born in Goshen, Orange County, New York, on July 8, 1810, and began the practice of law in 1837 in Angelica, Alleghany County, New York. By the end of 1837, however, he and W. W. Murphy set up a practice in Jonesville, Michigan. Judge Howell's political career began in the early 1840s when he was

appointed as district attorney for Hillsdale County. He later served in the Michigan state senate between 1843 and 1846. He then apparently returned to private practice, where he was identified with such causes as the abolition of capital punishment, establishment of free public schools, and the right of married women to hold property in their own name. He then returned to public service in 1859 as the prosecuting attorney for the newly created Mecosta County, Michigan, and served in the Michigan legislature between 1861 and 1863. The suggestion that Judge Howell should receive a federal appointment came by way of a petition to President Lincoln on March 20, 1862, from the Michigan delegation asking that Howell be made a district attorney or a judge in some western territory. This petition was approved on March 10, 1863, when Howell was given a four-year commission as judge in the newly created Arizona Territory. See Goff, "William T. Howell," 221.

5. The First Judicial District consisted of the area south of the Gila River and east of the 114th degree longitude, which under the districts established for the First Legislative Session, would become Pima County. See Walker and Bufkin, *Historical Atlas of Arizona*, 29.

6. Arizona Legislative Journals, 1st Territorial Assembly, 33–34 (1864).

7. See Browne and Taylor, *Reports upon Mineral Resources*, 247–48; and Yale, *Legal Titles to Mining Claims*, 84–85.

8. Arizona Legislative Journals, 1st Territorial Assembly, 70–71 (1864).

9. Arizona Legislative Journals, 1st Territorial Assembly, 130–31 (1864). Arizona Legislative Journals, 1st Territorial Assembly, 178–79 (1864).

10. Arizona Legislative Journals, 1st Territorial Assembly, 192 (1864).

11. Arizona Legislative Journals, 1st Territorial Assembly, 192 (1864).

12. This was in marked contrast to the regulations of the Walker Mining District; see chapter 2.

13. Arizona Legislative Journals, 1st Territorial Assembly, 205–6 (1864); Arizona Laws, 1st Territorial Legislature, xi (1864).

14. Letter to William A. Richmond. See Goff, "William T. Howell," 223.

15. Judge Howell's effort was published by the *Arizona Miner* in 1865 and is hereafter referred to as the Howell Code. The mining code is codified as Chapter L, 391–404.

16. Bill of Rights, article 21, Howell Code, 20.

17. The language from the bill of rights is apparent from limited editing of sections 1 and 2 of chapter V of the *Ordenanzas de Mineria* which, in Rockwell, *Spanish and Mexican Law*, 49, is translated as follows: "The mines are the property of my Royal Crown, . . . without separating them from my Royal patrimony, I grant them to my subjects in property and possession, in such manner that they may sell, exchange (pass by will, either in the way of inheritance or legacy), or in any other manner, dispose of all their property in them, upon the terms on which they themselves possess it, and persons legally capable of acquiring it."

18. Chapter 50, section 5, Howell Code.

19. Chapter 50, section 12, Howell Code.

20. The term *pertenencia* seems clearly a derivation of a term used by the miners

in the Middle Ages. In the memorialization of the Inquisition at Ashbourne in 1288, ordered by Edward I of England to determine the practices of the miners of the Peak District, a "meer" was defined as containing four *perticatas*, which although the word might have correctly been *pertica* nonetheless uses the same root. See Gordon, "Quo Warranto," 219–21.

21. This language suggests that the classic right of extralateral pursuit outside the boundaries of a claim was limited to only the extent of the vein within the limits of the surface boundaries of the claim.

22. Chapter 50, section 15, Howell Code.

23. Chapter 50, title 5, section 52, Howell Code.

24. Chapter 50, section 19, Howell Code.

25. Chapter 50, section 20, Howell Code.

26. *American Law of Mining*, 2nd ed., vol. 3, chap. 60.

27. Chapter 50, section 51, Howell Code. Colorado later adopted a similar provision requiring the original locator to stake one hundred feet on either side of the discovery claim for the common schools. Act of February 9, 1866.

28. English v. Johnson, 17 Cal. 107 (1860).

29. See Union Oil Co. of California v. Smith, 249 U.S. 337 (1919); Ranchers Exploration Co. v. Anaconda Co., 247 F. Supp. 705 (D. Utah 1965); MacGuire v. Sturgis, 347 F.Supp. 580 (D. Wyo. 1971); Geomet v. Lucky Mc Uranium Corp., 124 Ariz. 55, 601 P.2d 1339 (1979), cert. dismissed, 448 U.S. 917 (1980).

30. See title VI, article 2 of *Ordenanzas de Mineria.*

31. Chapter 50, section 23, Howell Code.

32. Chapter 50, section 24, Howell Code.

33. Chapter 50, section 36, Howell Code.

34. Chapter 50, section 34, Howell Code.

35. Chapter 50, section 35, Howell Code.

36. Chapter 50, section 28, Howell Code.

37. Chapter 50, section 29, Howell Code.

38. Chapter 50, section 32, Howell Code.

39. Chapter 50, section 18, Howell Code.

40. Chapter 50, section 31, Howell Code.

41. William Howell, letter to William A. Richmond, March 9, 1864, in Goff, "William T. Howell," 232; Farish, *History of Arizona*, 3:150–51n27: "A silver mine once opened and tested is a safe and permanent investment. Mines are discovered, tested, found rich and stock immediately created and sold in such a diluted condition that the investment is unprofitable."

42. Chapter 50, section 32, Howell Code. This is the precise procedure later adopted by the US Congress in the General Mining Law. This provision, codified as United States Code, Title 30, § 28c, reads: "The period for which said deferment may be granted shall end when the conditions justifying deferment have been removed: Provided, That the initial period shall not exceed one year but may be renewed for a further period of one year if justifiable conditions exist: . . ."

43. Chapter 50, section 37, Howell Code. A judicial definition of denouncement in the decisions of the courts of this country can be found in Steward v. King, 166 Pac. 55, 56 (Ore. 1917) (interpreting the laws of Panama).

44. Chapter 50, section 40, Howell Code.

45. Chapter 50, section 25, Howell Code.

46. Given the state of modern communications, it is ironic that a similar provision requiring the filing of mining claim notices with the Bureau of Land Management under the Federal Land Policy and Management Act of 1976 allowed three years for such action. United States Code, Title 43, § 1744. This suggests either that the old-time miners were much more diligent in the maintenance of their rights or today's legislators think that the miners need a substantially longer time to consult with their lawyers. As a practical matter, most miners simply put off the filing until the last minute, and a six-month period would probably have had the same effect. Chapter 50, sections 25.2 and 26, Howell Code.

47. Chapter 50, section 25.3, Howell Code.

48. Chapter 50, section 27, Howell Code. The Third Judicial District, i.e., Yavapai County as originally constituted, may have been the only area to remotely attempt to comply with this provision, as the Yavapai County Recorder has the original district books of the Pioneer, Walker Quartz, Quartz Mountain, Yavapai, and Turkey Creek Districts.

49. Chapter 50, section 45, Howell Code.

50. Chapter 50, section 48, Howell Code.

51. The Statute of Anne enacted by the British Parliament in 1705 changed the common law and imposed liability of co-owners in developing mineral property. Partition of mineral estates is something that can rarely be done because of the problem of appraising segregated portions of the land. See McCord v. Oakland Quicksilver Mining Co., 64 Cal. 134, 27 Pac. 863 (1883), where the California Supreme Court suggested that "the uncertainty of the prospective value of the [mining] property may render it impossible to make a just partition . . ." Curtis Lindley was also of the same opinion, see Lindley, *Treatise on the American Law*, vol. 3, § 792, 1953–54 ("in the case of lodes and veins, it would seem impossible to effect a fair, actual division").

52. Chapter 50, section 50, Howell Code.

53. Act of July 26, 1866, 14 Stat. 252.

54. *Congressional Globe*, 32d Cong., 1st Sess., App. 18 (1851).

55. The bill would grant rights to an "actual settler, to enter upon and remain on any public land of the United States containing minerals, not specifically reserved for public uses, within the States of California and Oregon, and to work the mines on the said lands for their own use and benefit, according to the laws and usages of the said States, respectively, and no person who has heretofore worked the said mines on said lands for their own use and benefit shall be regarded as a trespasser against the United States . . ." Yale, *Legal Titles to Mining Claims*, 347.

56. Ehrenberg, "Letter to the Editor," 202–3.

57. Ehrenberg, "Letter to the Editor," 211–12.

58. Indeed, after the Union capture of Tucson on May 20, 1862, 1st Lt. Edward

Banker Willis proceeded to arrest Sylvester Mowry at his mine. Charged with selling lead to Arizona's Confederate militias, Mowry was detained at Fort Yuma from July 2, 1862, to November 8, 1862. He was released after a court appearance in which no evidence could be found that he had ever sold lead to rebels.

59. Mowry, *Arizona & Sonora*, 202.

60. Mowry, *Arizona & Sonora*, 203.

61. Mowry, *Arizona & Sonora*, 204.

62. Mowry, *Arizona & Sonora*, 209.

63. In spite of claims that the remarks constituted a threat of secession (which is not surprising in the context of Mowry's politics in 1864), his warning was not without historical precedent. Eleven years earlier gold discoveries near Ballarat in southern Victoria, Australia, produced another wave of miners in search of the newest El Dorado. The Australian colonial government, in contrast to the circumstances found in California four years earlier, was organized and immediately improvised a program of licensing of miners. However, abuses and disparity of application of this system provoked a revolt among the miners that culminated with the burning of the Eureka Hotel in a riot and a bloody battle where miners were crushed by colonial troops at what would be known as the "Eureka Stockade." See Anderson, *Eureka, Victorian Parliamentary Papers*. Had the same chain of events occurred in early California, the sparse federal troop garrison would most certainly not have had the capability of enforcing a similar result. There was, however, considerable sympathy for Mowry's views within the Congress, and the western representatives pointed out during the debates that a futile attempt had been made in 1835 to drive miners out of the Missouri lead fields by federal troops. It is also likely that the political fallout over President Lincoln's order of seizure of the New Almaden Quicksilver Mine in 1863 was still relatively fresh in the minds of the members of Congress.

64. Mowry, *Arizona & Sonora*, 213.

65. Mowry, *Arizona & Sonora*.

66. Mowry, *Arizona & Sonora*.

67. Mowry, *Arizona & Sonora*, 215.

68. Lownsdale v. Parrish, 62 U.S. (21 How.) 290 (1859).

69. Act of February 27, 1865, 13 Stat. 440.

70. Libecap, "Government Support of Private Claims," 371. The potential availability of land was also addressed in article IV of the treaty of January 17, 1865, concluded on October 12, 1863, between United States and Shoshone-Goship Bands of Indians, Tuilla Valley, Utah (Montana), "It is further agreed by the parties hereto that the country of the Goship tribe may be explored and prospected for gold and silver, and other minerals and metals; and when mines are discovered, they may be worked, and mining and agricultural settlements formed, and ranchos established wherever they may be required." 13 Stat. 682. See also Yale, *Legal Titles to Mining Claims*, 349.

71. H.R. Misc. Doc. No. 14, 38th Cong., 2d Sess. (January 25, 1865).

72. *Arizona Miner*, June 13, 1866, 4:4.

73. 14 Stat. 43 (1866).

74. King, *United States Mining Laws*, 330, 508.

75. See letter of E. F. Dunne in Raymond, *Statistics of Mines and Mining* (1870), 421–22.

76. Yale was a well-known mining lawyer at the time and the author of *Legal Titles to Mining Claims and Water Rights in California* (1867). He came to California in 1850 and established a law practice in San Francisco where for a number of years he was associated with the law firm of Halleck, Peachy & Billings. Yale, *Legal Titles to Mining Claims*, 9–10.

77. Yale, *Legal Titles to Mining Claims*, 12.

78. Raymond, *Statistics of Mines and Mining* (1870), 426.

79. Raymond, *Statistics of Mines and Mining* (1870), 422.

80. Raymond, *Statistics of Mines and Mining* (1870), 421.

81. Raymond, *Statistics of Mines and Mining* (1870), 423.

82. See Thomas v. Union Pacific Railroad Company, 139 F.Supp. 588 (D.Colo. 1956).

83. *Congressional Globe*, 42nd Cong., 2nd Sess. 532 (1872).

84. Apologies to purists who understand that a "tunnel" requires both an entrance and an exit. An "adit" would be more appropriate—under Spanish law this would be a *boca mina.*

85. The law then went on to require the performance of annual assessment work of one hundred dollars per year and include a procedure to force a transfer of ownership if co-owners did not contribute to these requirements.

Chapter 4: Revival and Operation of the Arizona Mining District Laws

1. Act of July 26, 1866, 14 Stat. 252.

2. *Arizona Miner*, 2:5, 6 (Sept. 12, 1866) and 1:5 (Sept. 26, 1866).

3. *Arizona Miner*, 2:2 (Oct. 13, 1866).

4. Browne and Taylor, *Reports upon Mineral Resources* , 136–37.

5. Browne and Taylor, *Reports upon Mineral Resources*, 137.

6. Browne and Taylor, *Reports upon Mineral Resources*, 137.

7. *Daily Arizona Miner* (Oct. 26, 1866).

8. *Arizona Miner* 1:3 (Oct. 13, 1866).

9. Act of July 26, 1866, 14 Stat. 252.

10. Arizona Laws, 3rd Territorial Assembly (1866), chap. 13, § 1; Arizona Compiled Laws, 1864–1871, chapter 50, section 1. One can assume that this provision may have been directed to some rather loose practices. For example, Frank Love provides an 1863 description of recorder Jacob Snively's office at the Castle Dome District as "a battered valise, a box labeled 'Olive Oil,'' which functioned as a desk, a pot, frying plan, shotgun, and several blankets." *Daily Alta California*, January 2, 1864, in Love, *Mining Camps and Ghost Towns*, 3. On the other hand, the mining district records eventually on file with the Yavapai County Recorders had been kept with some care.

11. Arizona Compiled Laws, 1864–1871, chapter 50, section 2.

12. Arizona Compiled Laws, 1864–1871, chapter 50, section 4.

13. Arizona Compiled Laws, 1864–1871, chapter 50, section 5.

14. See Pine Grove Mining District Regulations, *Arizona Weekly Miner*, 1:1 (Feb. 25, 1871); Bradshaw Mountain Mining District, King, *United States Mining Laws*, 264–65.

15. Pine Grove Mining District Regulations, Jan. 31, 1871, *Arizona Weekly Miner*, 1:1 (Feb. 25, 1871).

16. This is the only suggestion that any nonmetallic mineral might be subject to location in any of the district regulations, although this reference may only refer to a host material.

17. Bradshaw Mountain Mining District Regulations, King, *United States Mining Laws*, 264–65.

18. Art. 3d, Fort Rock Mining District Regulations, Jan. 27, 1872, *Arizona Weekly Miner*, 3:3 (Feb. 17, 1872).

19. Tiger Mining District Regulations, Feb. 4, 1871, *Arizona Weekly Miner*, 3:2 (May 13, 1871), King, *United States Mining Laws*, 269; Sedgwick Mining District Regulations, Nov. 20, 1871, *Arizona Weekly Miner*, 3:2 (Jan. 1, 1872), King, *United States Mining Laws*, 270.

20. Fort Rock Mining District Regulations, Jan. 27, 1872, *Arizona Weekly Miner*, 3:3 (Feb. 17, 1872). Regulations of Bradshaw Mountain Mining District.

21. Wallapai Mining District Regulations, Nov. 4, 1870, King, *United States Mining Laws*, 268.

22. Regulations of the Copper Mountain Mining District, Feb. 10, 1872, Arizona State Archives.

23. Art. 2, Wallapai Mining District Regulations, Mar. 22, 1871, King, *United States Mining Laws*, 269 (emphasis in original).

24. Dunning and Peplow, *Rock to Riches*, 69–70.

25. Dunning and Peplow, *Rock to Riches*, 96; Hinton, *Handbook to Arizona*, 72–167.

26. Revised Statutes of the United States § 2324 (1876), United States Code, Title 30, § 28.

27. Silver Mining District Regulations, Jan. 20, 1879, Book 1 (Misc.), pages 303–4, Yuma County.

28. Sec. 7, Truman Mining District Regulations, Feb. 11, 1875, Book 1 (Misc.), page 296; *Arizona Citizen*, 4:2, 3 (Mar. 13, 1876).

29. Sec. IV, Warm Spring Mining District Regulations, Book 1 (Promiscuous), pages 80–81, Coconino County.

30. Article Second, Pioneer Mining District Regulations, July 11, 1875, Book 1 (Misc.), page 9, Pinal County Records, *The Arizona Citizen*, 1:2 (Aug. 10, 1875).

31. Cedar Valley Mining District Regulations, Sept. 10, 1874, Arizona State Archives.

32. Smith Mining District Regulations, Apr. 3, 1875, *The Tucson Daily Citizen*, 4:2 (Mar. 17, 1875).

33. California Mining District Regulations, Sept. 7, 1878, Arizona State Archives.

34. Richardson Mining District Regulations, Nov. 18, 1882, Book 1 (Misc.), pages 614–18, Mohave County.

35. Section 10, Greaterville Mining District Regulations, Nov. 7, 1880, Arizona Historical Society Records.

36. Article V, California Mining District Regulations, Sept. 7, 1878, Arizona State Archives.

37. Wrightson Mining District Regulations, Mar. 13, 1881, *The Arizona Citizen* (Mar. 22, 1881).

38. Greenlee Gold Mountain Mining District Regulations, Apr. 28, 1879, Arizona State Archives.

39. Sec. 5, Aztec Mining District Regulations, May 9, 1877, Book 1 (Misc.), page 449, Pima County; Art. 4, Pajarita Mining District Regulations, June 5, 1880, Book 2 (Misc.), pages 212–16, Pima County.

40. For example, two dollars was required by Pima, Plomosa, Pajarita and Helvetia; Cedar Valley required one dollar; and Richardson and Minnesota required three dollars.

41. Art. 12, Pajarita Mining District Regulations.

42. Ninth Article, Verdi Mining District Regulations, Apr. 17, 1876, Yavapai County Records.

43. Sec. 13, Smith Mining District Regulations.

44. Section III, Warm Spring Mining District Regulations, Book 1 (Promiscuous), pages 80–81, Coconino County.

45. Five dollars: see Pima, Little Cottonwood, Arivaca and Hartford Mining District Regulations; six dollars: see Greenlee Gold Mountain, Cochise and Yellow Stone Mining District Regulations.

46. Plomosa Mining District Regulations, Feb. 11, 1879, *Arizona Sentinel*, 2:1 (April 5, 1879). See also Richardson Mining District Regulations, Nov. 18, 1882, Book 1 (Misc.), pages 614–18, Mohave County.

47. Section 10, Tyndall Mining District Regulations, Nov. 17. 1876, Book 1 (Misc.), pages 394–97, Pima County.

48. Pajarita Mining District Regulations, article 12.

49. Pajarita Mining District Regulations, article 14.

50. Seattle Mining District Regulations, Jan. 1, 1881, Arizona Historical Society Records; Article 7; Greaterville Mining District Regulations, *The Tucson Daily Citizen*, resolution Jan. 1, 1881, section 18 (March 17, 1881); Globe Mining District Regulations, Nov. 25, 1875, *Arizona Weekly Miner*, 1:3 (Dec. 17, 1875).

51. Pajarita Mining District Regulations, article 12.

52. Smith Mining District Regulations, section 10.

53. Silver Mining District Regulations, Jan. 20, 1879, Book 1 (Misc.), pages 303–4, Yuma County.

54. Richardson Mining District Regulations.

55. Smith Mining District Regulations, sec. 7.

56. Sec. 11, Greaterville Mining District Regulations, Nov. 27, 1880, Arizona Historical Society Archives.

57. Sec. 13, Pajarita Mining District Regulations, June 5, 1880, Book 2 (Misc.), pages 212–16, Pima County. Sec. II, Warm Spring Mining District Regulations, Apr. 4, 1896, Book 1 (Promiscuous), pages 89–90, Coconino County.

58. *Report of the Public Lands Commission*, 646.

59. Browne and Taylor, *Reports upon Mineral Resources*, 229.

60. Records of proceedings held Oct. 17, 1865, San Francisco Mining District, 1–3, Mohave County Historical Society.

61. Raymond, *Statistics of Mines and Mining* (1874), 513.

62. Act of March 3, 1879, 20 Stat. 394.

63. *Report of the Public Lands Commission*, 396.

64. *Report of the Public Lands Commission*, 398–99.

65. See Article 2, Harshaw Mining District Regulations, Apr. 29, 1878, Book 1 (Misc.), page 545, Pima County; Tucson Mining District Regulations, Apr. 30, 1878, Book 1 (Misc.), page 548, Pima County.

66. See comments in *Report of the Public Lands Commission*, by J. D. Hyde, Tulare County, California, Oct. 28, 1879, 78; J. B. Low, Denver, Colorado, Sept. 2, 1879, 291; N. Armstrong, Beaver Head County Montana, September 28, 1879, 340; Orville B. O'Bannon, Deer Lodge County, Montana, Sept. 28, 1879, 341; Walter McDermott, Lewis & Clarke County, Montana, 373; and Harvey Carpenter and James McMartin, Eureka, Nevada, Nov. 3, 1879, 424.

67. *Report of the Public Lands Commission*, 85.

68. *Report of the Public Lands Commission*, 81 at 85–86.

69. Elliott, *Servant of Power*, 67–68.

70. *Report of the Public Lands Commission*, 202.

71. Raymond, *Mineral Resources* (1869), 221–22.

72. *Report of the Public Lands Commission*, 646.

73. *Report of the Public Lands Commission*, lxxix, Section 174 (third).

74. Colorado Laws, 10th Territorial Assembly, 185–93 (1874). These laws required lode claims to be not more than 1,500 feet in length and 150 feet on each side of the vein; three months to record a location notice; sixty days to sink a 10-foot location shaft; and six monuments, one at each corner and one at the center of each end line.

75. *Arizona Miner*, 1:5 (Sept. 21, 1867).

76. Arizona Laws, 18th Territorial Assembly, chapter 42 (1895).

77. See Dunning and Peplow, *Rock to Riches*, 67–127.

78. See "District Mining Regulations," in Hinton, *Handbook to Arizona*, appendices vi–viii; "Form of District Rules and Organization," in Wilson, *Manual of the Mining Laws*, 95–98; and Morrison, *Mining Rights in Colorado* (1874), 4–8. Morrison's guide, published in 1874, includes some interesting guidance in suggesting "the mode by which dueling should be conducted; bounties on married women; penalties for rolling rocks down the mountain, etc., are among the matters often provided for." Morrison, *Mining Rights in Colorado* (1874), 4–8.

79. See Copp, *Manual for the Use*.

80. Arizona Laws, 20th Territorial Assembly, chapter 14 (1899).

81. Book 4 (Misc.), pages 533–38.

82. See Superstition Mining District Records, sections 5, 7, and 8.

83. Book 1 (Promiscuous), pages 89–90, Coconino County.

84. Hill, *Mining Districts*; Wilson et al., *Map and Index of Arizona*.

85. Arizona Laws, 1st Territorial Assembly, 511 (1864).

86. Arizona Laws, 4th Territorial Assembly, 513 (1867).

87. Howell Code, chapters xxxiii (sec. 14) and chapter li (sec. 24).

88. Arizona Laws, 8th Territorial Assembly, p. 333 (1875); Arizona Compiled Laws, Sections 2062–67 (1877). Gov. A. P. K. Safford argued that this was not a tax on mines but on capitalists. The history of mine taxation in Arizona has been described in 1944 as having run full cycle, "with a cumulative drive for higher taxes as the mines prospered, and a cumulative decline as they became depleted." See Roberts, *State Taxation of Metallic Deposits*, 189–274. For a general description of the nature of mine taxation, see Schenck, *Handbook of State and Local Taxation.*

89. Arizona Compiled Laws, section 2072 (1877).

90. Arizona Compiled Laws, section 2072 (1877).

91. Arizona Laws, 11th Territorial Assembly, chapter 95, 167 (1881).

92. Arizona Laws, 11th Territorial Assembly, chapter 99, 171 (1881).

93. Revised Statutes of Arizona, para. 2352–57, 412–13 (1887).

94. Arizona Laws, 18th Territorial Assembly, chapter 26, 32 (1895).

95. Arizona Laws, 19th Territorial Assembly, chapter 59, 111 (1897).

96. Arizona Laws, 25th Territorial Assembly, chapter 60 (1909).

97. See United States Code, Title 43, § 291 et. seq. (Stockraising Homestead Act of 1916); United States Code, Title 43, § 399(g) (exchange provisions of Taylor Grazing Act of 1934).

98. Arizona Laws, 26th Legislature, 2nd Regular Session, chapter 34, § 1 (1964).

99. SMEA was organized in the mid-1960s by a group of individual geologists who called themselves the "Dirty Dozen" for the purpose of writing replies to antimining newspaper articles and editorials in the Tucson newspaper. The organization took on an air of legitimacy when it changed its name in 1971 and began active lobbying efforts to improve the Arizona mining laws (although it was still viewed with suspicion by the major mining industry because of its association with some environmental groups).

100. Arizona Laws, 33rd Legislature, 2nd Regular Session, chap. 177 (1978).

101. Section 314, Federal Land Policy and Management Act of 1976, United States Code, Title 43, § 1744; *Federal Register*, vol. 42, 5298–5300 (Jan. 27, 1977).

102. Arizona Laws, 33rd Legislature, 2nd Regular Session, chap. 177, § 8 (1978); Arizona Revised Statutes § 27–210.

103. The wording of the act was negotiated at a spring 1975 meeting between representatives of the Arizona Mining Association, represented by Howard Twitty, and Southwestern Minerals Exploration Association.

104. Arizona Laws, 39th Legislature, 1st Regular Session, chapter 249, § 10 (1989).

105. Arizona Revised Statutes, § 27–205.

106. *Arizona Miner*, 1:5 (Jan. 26, 1867).

107. *Arizona Daily Miner*, 2:1 (Dec. 28, 1874).

108. Arizona Revised Statutes, § 27–209 ("it shall be a sufficient description of a mining claim if the name of the claim, the district, county and state where it is located, and the book and page where the location notice thereof is recorded can be understood therefrom."); Hill, *Mining Districts*, 94. The map included as an illustra-

tion made a general attempt to define the boundaries of these various districts, but these lines can only be considered illustrative because boundaries frequently changed as a result of adjustments to accommodate the formation of new districts, disuse, and reorganization under new names.

Appendix B: Full Text of Arizona Mining District Regulations

1. Rynerson was a former California Volunteer, southern New Mexico politician, and mine speculator.

2. Joy was a civil engineer working for Eber Ward of Detroit in charge of securing the future Morenci Mine ground.

3. There is much on Metcalfe in Conger and Cogut, *History of Arizona's Clifton-Morenci Mining District* (1999).

4. There is much on the Freudenthals in Elizabeth L. Romenofsky, *From Charcoal to Banking, the I.E. Solomons of Arizona* (1984).

5. Businessman and mayor of Tucson.

6. Also D. B. Rae since he used both. Attorney David Baxter Rea (1836–1899) used "Rea" mostly while in Arizona, but in his native state of North Carolina he used Rae, as he did in California and elsewhere. According to the end of the secretary's notes here and the April 3, 1875, *Citizen*, he was author of the Smith district bylaws.

7. In response to the questioning of the validity of the Tombstone district formation by Judge Wells Spicer, miner Tom Walker (Richard Gird's relative) wrote, "I flatly contradict it, and reply that we have a district in Tombstone; properly established according to law, and the boundaries already defined. One year ago, last April we signed a call; posted it; organized and adopted rules and regulations." T. E. Walker, "Tombstone Items," *Arizona Citizen*, Tucson, November 28, 1879, 2:1; reprint from Tombstone *Nugget*.

Bibliography

Books and Articles

Aiton, Arthur S. "The First American Mining Code." *Michigan Law Review* 23 (1924).

Aiton, Arthur S. "Ordenancas Hechas por el Sr. Visorrey Don Antonio de Mendoca Sobre las Minas de la Nueva España Año de M.D.L." *Revista de Historia de America* (June 1942). Special Collections, University of Arizona Library.

Aiton, Arthur Scott. *Antonio de Mendoza, First Viceroy of New Spain*. Duke University Press, 1927.

American Law of Mining. 2nd ed. Lexis/Nexis-Matthew Bender & Co., 1984.

Anderson, Hugh, comp. *Eureka, Victorian Parliamentary Papers, Votes and Proceedings 1854–1867*. Hill of Content, 1969.

Bakken, Gordon M. *The Development of Law on the Rocky Mountain Frontier: Civil Law and Society, 1850–1912*. Greenwood Press, 1983.

Bancroft, Hubert H. *History of Arizona and New Mexico*. The History Company, 1890.

Bancroft, Hubert H. *History of North Mexican States*. The History Company, 1883.

Bartlett, Katherine. "Notes upon the Routes of Espejo and Farfan to the Mines in the Sixteenth Century." *New Mexico Historical Review* 17, no. 1 (1942).

Beers, Henry P. *Spanish and Mexican Records*. University of Arizona Press, 1979.

Blake, William Phipps. *Diary*. Arizona Historical Society, Tucson, Arizona, Papers 1847–1910, MS 78.

Bolton, Herbert E., ed. *Spanish Exploration in the Southwest 1542–1706*. Barnes & Noble, 1908.

Bowden, J. J. "Private Land Claims in the Southwest, The Ortiz Mine Grant." Master's thesis, University of Houston, 1969.

Brown, George Rothwell, ed. *Reminiscences of Senator William M. Stewart of Nevada*. The Neale Publishing Co., 1908.

Browne, J. Ross. *Report on the Mineral Resources of the States and Territories West of the Rocky Mountains*. Government Printing Office, 1868.

Browne, J. Ross. *A Tour Through Arizona 1864*. Originally published in serial form in *Harper's*

News Magazine in 1864. Republished, Arizona Silhouettes, 1951.

Burggraaf, Pieter. *Across New Mexico and Arizona Territories and up the Hassayampa River, 1861–1862, The Revised Story*. CreateSpace, 2015.

Christiansen, Paige W. *The Story of Mining in New Mexico*. New Mexico Bureau of Mines & Mineral Resources, 1974.

Clissold, Stephen. *The Seven Cities of Cíbola*. Clarkson N. Potter, 1962.

Colección de documentos inéditos relatives al descubrimiento, conquista y organización de las antiquas posesiones españolas de América y Oceanía, sacados de los archivos del Reino y muy especialmente de las Indias, series 1, 10:40 and 26:67. University of Florida Digital Collections. https://ufdc.ufl.edu/AA00037906/00039/images.

Conger, Bill, and Ted Cogut. *History of Arizona's Clifton-Morenci Mining District : A Personal Approach*. Mining Hi-Story, 1999.

Conner, Daniel Ellis. *Joseph Reddeford Walker and the Arizona Adventure*. Edited by Donald J. Berthrong and Odessa Davenport. University of Oklahoma Press, 1956.

Copp, Henry N. *Manual for the Use of Prospectors on the Mineral Lands of the United States*. 5th ed. Copp, 1897.

Cozzens, Samuel Woodworth. *The Marvelous County; or Three Years in Arizona and New Mexico, The Apaches' Home*. Facsimile reprint. Ross & Haines, 1967.

Crombie, Katherine, Chris T. Gholson, Dante S. Lauretta, and Erik B. Melchiorre. *Rich Hill, the History of Arizona's Most Amazing Gold District*. Golden Retriever Publications, 2002.

Dawson, Joseph G., III. *Doniphan's Epic March, The 1st Missouri Volunteers in the Mexican War*. University Press of Kansas, 1999.

DeQuille, Dan. *The Big Bonanza*. Originally published 1876. Republished, Alfred Knopf, 1947.

Dunning, Charles H., and Edward H. Peplow Jr. *Rock to Riches*. Southwest Publishing Co., 1959.

Ehrenberg, Herman. "Letter to the Editor." *Mining Magazine and Journal of Geology* (December 1859): 202–11.

Elliott, Russell R. *Servant of Power: A Political Biography of Senator William M. Stewart*. University of Nevada Press, 1983.

Ely, James W., Jr. *The Guardian of Every Other Right: A Constitutional History of Property Rights*. Oxford University Press, 1992.

Etter, Patricia A. *To California on the Southern Route 1849: A History and Annotated Bibliography*. Arthur H. Clarke Company, 1998.

Fedner, Lawrence T. *Fort Wilkins 1844 and the U.S. Mineral Land Agency 1843, Copper Harbor, Michigan, Lake Superior*. Vantage Press, 1966.

Flint, Richard. *No Settlement, No Conquest: A History of the Coronado Entrada*. University of New Mexico Press, 2008.

Goff, John S. "William T. Howell and the Howell Code of Arizona." *American Journal of Legislative History* 11 (1967): 221–33.

Gómez, Laura E. *Manifest Destinies: The Making of the Mexican American Race*. New York University Press, 2007.

Gordon, David. "The Quo Warranto—A New Translation." *Bulletin of the Peak District Historical Society* 10 (1988): 219–23.

Gray, Andrew B. *Survey of a Route for the Southern Pacific Railroad on the 32nd Parallel*. Wrightson & Co., 1856. Reprinted, edited by L. R. Bailey. Westernlore Press, 1963, including "The Reminiscences of Peter R. Brady of the A. B. Gray Railroad Survey, 1853–1854."

Hamilton, Patrick. *The Resources of Arizona*. 3rd ed. Bancroft, 1884.

Hendricks, Rick. "Spanish Colonial Mining in Southern New Mexico: A Spanish to English Translation of Documents relating to El Paso, the Organ Mountains and Santa Rita del Cobre." *Mining History Journal* 6 (1999): 143–62.

Hinton, Harwood. "Frontier Speculation: A Study of the Walker Mining Districts." *Pacific Historical Review* 29 (1960): 245.

Hinton, Richard J. *The Handbook to Arizona: Its Resources, History, Towns, Mines, Ruins and Scenery*. Payot, Upham & Co., and American News Co., 1878. Republished, Rio Grande Press, 1970.

Huggard, Christopher J., and Terrence M. Humble. *Santa Rita del Cobre, a Copper Mining Community in New Mexico*. University Press of Colorado, 2012.

Hunt, Aurora. *Kirby Benedict, Frontier Federal Judge*. Arthur H. Clark, 1961.

Hurst, James Willard. *Law and Economic Growth: The Legal History of the Lumber Industry in Wisconsin, 1836–1915*. Belknap Press of Harvard University Press, 1964.

Jones, Fayette A. *New Mexico Mines and Minerals*. Santa Fe, 1905. Facsimile reprint as *Old Mines and Ghost Camps of New Mexico*, Frontier Book Co., 1970.

Kessell, John L., ed. *Remote Beyond Compare: Letter of Don Diego de Vargas to His Family from New Spain and New Mexico, 1675–1706*. University of New Mexico Press, 1989.

Lacy, John. "Going with the Current: The Genesis of the Mineral Laws of the United States." *41st Annual Proceedings of the Rocky Mountain Mineral Law Foundation*, paper 10 (1995): 1–53.

Lacy, John. "The Least Welcome Profession; Lawyers in the Mining Camps." *Dips, Angles & Spurs: Newsletter of the Mining Law Antiquarians*, March 1, 1982, www.mininglawhistory.org.

Lacy, John C. "Historical Overview of the Mining Law; the Miners' Law Becomes Law." In *The Mining Law of 1872: A Legal and Historical Analysis*. National Legal Center for the Public Interest, 1989.

Lamar, Howard Roberts. *The Far Southwest 1846–1912: A Territorial History*. W. W. Norton & Co., 1970.

Langley, Harold D., ed. Letter dated April 5, 1869. *To Utah with the Dragoons and Life in Arizona and California 1858–1859*. University of Utah Press, 1974.

Larson, Robert W. *New Mexico's Quest for Statehood*. University of New Mexico Press, 1968.

Libecap, Gary D. "Economic Variables and the Development of the Law: The Case of Western Mineral Rights." *The Journal of Economic History* 38 (June 1978): 338–62.

Libecap, Gary D. "Government Support of Private Claims to Public Mineral: Western Mineral Rights." *Business History Review* 53 (1979): 364.

Lindley, Curtis H. *A Treatise on the American Law Relating to Mines and Mineral Lands Within the Public Land States and Territories*. 3rd ed. Bancroft-Whitney, 1914.

Love, Frank. *Mining Camps and Ghost Towns*. Westernlore Press, 1974.

Melchiorre, Erik. *Gold Atlas of Rich Hill, Arizona*. Rock Doc Publications, 2009.

Meketa, Jacqueline Dorgan, ed. *Legacy of Honor: The Life of Rafael Chacón, a Nineteenth-Century New Mexican*. University of New Mexico Press, 1986.

Morrison, Robert S. *Mining Rights in Colorado*. 7th ed. Chain & Hardy Co., 1894.

Morrison, Robert S. *Mining Rights in Colorado*. Rocky Mountain News, 1874.

Mowry, Sylvester. *Arizona & Sonora: The Geography, History and Resources of the Silver Region of North America*. 3rd ed. Harper & Brothers, 1864.

Mumey, Nolie. *Notes to Kearny Code*. Facsimile republication of Laws of the Territory of New Mexico, October 7, 1846, as presented by Oliver Perry Hovey, printer, to Col. A. W. Doniphan, 1970.

North, Diane M. T. *Samuel Peter Heintzelman and the Sonora Exploring and Mining Company*. University of Arizona Press, 1980.

Officer, James E. "Mining in Hispanic Arizona: Myth and Reality." *History of Mining in Arizona* 2, Mining Club of the Southwest Foundation, 1991.

Pagden, Anthony R., trans. and ed. *Hernán Cortés: Letters from Mexico*. Grossman, 1971.

Paul, Rodman Wilson. *Mining Frontiers of the Far West 1848–1880*. University of New Mexico Press, 1963.

Peplow, Edward. *History of Arizona*. Lewis, 1958.

Polzer, Charles W. "Legends of Lost Missions and Mines." *The Smoke Signal* 18, Tucson Corral of Westerners (1968): 170.

Poston, Charles D. *Building a State in Apache Land*. Aztec Press, 1963. Originally published by *Overland Monthly*, 1894.

Prescott, William A. *History of the Conquest of Peru*. J. B. Lippincott Co., 1873.

Pumpelly, Raphael. "Mineralogical Sketch of the Silver Mines of Arizona." *California Academy of Natural Sciences* 2 (1862): 138.

Pumpelly, Raphael. *My Reminiscences*. Henry Holt, 1918. Also published as *Across America and Asia*, Leypoldt & Holt, 1869. Reprinted, with extracted portions, including a commentary related to Arizona, as *Pumpelly's Arizona*, Palo Verde Press, 1965.

Quaife, Milo Milton, ed. *Christopher Carson, Kit Carson's Autobiography*. Lakeside Press, 1935. Reprint, University of Nebraska Press, 1966.

Reid, John Philip. *Constitutional History of the American Revolution: The Authority of Rights*. University of Wisconsin Press, 1986.

Roberts, Warren A. *State Taxation of Metallic Deposits*. Harvard University Press, 1944.

Rockwell, John A. *A Compilation of Spanish and Mexican Law in Relationship to Mines and Titles to Real Estate*. John S. Voorhies, 1851.

Ramenofsky, Elizabeth L. *From Charcoal To Banking, The I.E. Solomons of Arizona*. Westernlore Press, 1984.

Rose, Dan. *The Ancient Mines of Ajo*. Mission Press, 1936.

Schenck, George H. K. *Handbook of State and Local Taxation of Solid Minerals*. 2nd ed. Department of Mineral Economics of Pennsylvania State University, 1984.

Thompson, Charles. *The Ordinances of the Mines of New Spain*. London, 1825.

Thompson, Gerald. "Henry De Groot and the Colorado River Gold Rush, 1862." *Journal of Arizona History* (Summer 1996).

Twitchell, Ralph Emerson. *Leading Facts of New Mexican History*. Horn & Wallace, 1963. Reprint of 1911–12 original.

Wagner, Jay J. *Arizona Territory 1863–1912, A Political History*. University of Arizona Press, 1970.

Wagoner, Jay J. *Early Arizona, Prehistory to Civil War*. University of Arizona Press, 1975.

Walker, Henry P., and Don Bufkin. *Historical Atlas of Arizona*. University of Oklahoma Press, 1979.

Walker, Charles S. "Confederate Government in Doña Ana County: As Shown in the Records of the Probate Court, 1861–1862." *New Mexico Historical Review* (1931).

Wallace, Andrew, ed. *Pumpelly's Arizona, an Excerpt from Across America and Asia by Raphael Pumpelly, Comprising Those Chapters Which Concern the Southwest*. The Palo Verde Press, 1965.

Wilson, Charles S. *A Manual of the Mining Laws of the United States, Colorado, New Mexico and Arizona*. 2nd ed. W. H. Lawrence & Co., 1884.

Wilson, Robert Anderson. *Mexico and Its Religion*. Harper & Brothers, 1855.

Yale, Gregory. *Legal Titles to Mining Claims & Water Rights in California*. A. Roman & Co, 1867.

Judicial Decisions

William H. Hardy and Wooster M. Hardy Composing the Southern Cross Gold and Silver Mining Company vs. Whom It May Concern, in equity in probate court, Judicature of Mines, Mohave County, Arizona Territory. Library Archives of the Arizona Secretary of State (Phoenix).

Geomet Exploration Ltd. v. Lucky Mc Uranium Corporation, 124 Ariz. 60, 601 P.2d 1344 (App. 1979), appeal from Yuma County Cause No. C-38435 and C-38533; 124 Ariz. 55, 601 P.2d 1339 (1979); certiorari dismissed, 448 U.S. 917 (1980).

Lane v. Watts, 234 U.S. 525 (1914).

Lownsdale v. Parrish, 62 U.S. (21 How.) 290 (1859).

MacGuire v. Sturgis, 347 F.Supp. 580 (D.Wyo. 1971).

McCord v. Oakland Quicksilver Mining Co., 64 Cal. 134, 27 Pac. 863 (1883).

Union Oil Co. of California v. Smith, 249 U.S. 337 (1919).

United States v. Castillero, 67 U.S. (2 Black) 18 (1864).

Statutes and Codes

Act of May 20, 1862, chap. 75, 12 Stat. 392.

Act of July 26, 1866, 14 Stat. 252.

Arizona Compiled Laws, 1864–1871, Coles Bashford (Comp.) (Weed, Parsons and Company, 1871).

Arizona Compiled Laws, 1864–1877, John P. Hoyt (Comp.) (Richmond, Backus & Co., 1877).

Arizona Revised Statutes (Prescott Courier, 1887).

Arizona Revised Statutes (West Publishing Co., various dates).

Arizona Legislative Journals
- 1st Territorial Assembly (1864).
- 6th Territorial Assembly (1871).

Arizona Laws
- 1st Territorial Assembly (1864).
- 2nd Territorial Assembly (1865).
- 3rd Territorial Assembly (1866).
- 4th Territorial Assembly (1867).
- 8th Territorial Assembly (1875).
- 11th Territorial Assembly (1881).
- 18th Territorial Assembly (1895).
- 19th Territorial Assembly (1897).
- 20th Territorial Assembly (1899).
- 25th Territorial Assembly (1909).
- 33rd Legislature, 2d Sess., chap. 177 (1978).

Calif. Civil Code (1851) and (1872).

Howell Code (*Arizona Miner*, Prescott, 1865).

Laws of the Territory of New Mexico (Santa Fe, 1846).

New Mexico Territorial Laws (1851), H.R. Misc. Doc. 4.

Ordenanzas del Nuevo Cuaderno, codified as Law 9, Title 13, Book 6 of Recopilación de las Leyes de Estos Reynos (de Castilla) (Madrid: 1640).

Reales Ordenanzas para la Dirección, Régimen i Gobierno del Importante Cuerpo de la Minería de Nueva-España, i de su Real Tribunal General. De Orden de su Magestad (Madrid: 1783).

Recopilación de las Leyes de Estos Reynos (de Castilla) (Madrid: 1640). 3 volumes, specifically Title 13, Book 6, "De los Tesoros, i mineros de oro, o plata o otro qualquier metal, i pozos de sal, i bienes mostrencos, i hallados."

Revised Statutes of the United States (1876).

Statute of Anne, 4 & 5 Anne, chap. 16, § 27 (1705).

Treaty Between United States and Shoshone-Goship Bands of Indians, Tuilla Valley, Utah (Montana), concluded October 12, 1863.

United States Code (GPO, 1970).

Government Publications

Arizona Geological Survey. https://mininginfo.azgs.arizona.edu.

Blake, William P. *Report of the Governor of the Territory of Arizona to Secretary of the Interior for 1899*. Reprinted in *U.S. Geological Survey Professional Paper 209*, 98–99. Government Printing Office, 1946.

Browne, J. Ross, and James W. Taylor. *Reports upon Mineral Resources of the United States*. Government Printing Office, 1867.

Copp, Henry N. *United States Mineral Lands; Laws Governing Their Occupancy and Disposal; Decisions of Federal and State Courts in Cases Arising Thereunder; and Regulations and Rulings of the Land Department in Connection Therewith*. Copp, 1881.

Gray, Andrew B. "Report of the Commissioner of Indian Affairs, Accompanying the Annual Report of the Secretary of the Interior, for the Year 1859." https://babel.hathitrust.org/cgi/pt?id=mdp.39015019954125&seq=364&q1=%22A+B+Gray%22.

Donaldson, Thomas. *The Public Domain*. H.R. Misc. Doc. 45, 47th Cong., 2nd Sess. Government Printing Office, 1884.

Farish, Thomas Edwin. *History of Arizona*. Arizona Legislature, 1918, vol. 3.

Hill, James M. *The Mining Districts of the Western United States*. USGS Bulletin 507, Government Printing Office, 1912.

King, Clarence. *The United States Mining Laws and Regulations Thereunder and State and Territorial Mining Laws, to Which Are Appended Local Mining Rules and Regulations*. Report of the 10th United States Census. Government Printing Office, 1885.

Lord, Eliot. *Comstock Mining and Miners.* Government Printing Office, 1883.

Mowry, Sylvester. *Memoir of the Proposed Territory of Arizona.* Henry Polkinghorn, 1857. Reprint, Territorial Press, 1964.

Raymond, Rossiter W. *Mineral Resources of the States and Territories.* H.R. Ex. Doc. No. 54, 40th Cong., 3rd Sess. Government Printing Office, 1869.

Raymond, Rossiter W. *Statistics of Mines and Mining in the States and Territories West of the Rocky Mountains.* H.R. Exec. Doc. No. 207, 41st Cong., 2d Sess. Government Printing Office, 1870.

Raymond, Rossiter W. *Statistics of Mines and Mining in the States and Territories West of the Rocky Mountains.* Government Printing Office, 1874.

Report of New Mexico Convention of Delegates. H.R. Misc. Doc. No. 39, 31st Cong., 1st Sess., February 25, 1850.

Report of the Public Lands Commission Created by the Act of March 3, 1879, Relating to Public Lands in the Western Portion of the United States and to the Operation of Existing Land Laws. H.R. Exec. Doc. No. 46, 46th Cong., 2nd Sess. Government Printing Office, 1880.

Twitty, Sievwright, and Mills. "Nonfuel Mineral Resources of the Public Lands, Vols. 1–4, Legal Study." Public Land Law Review Commission, 30-1. National Technical Information Service, 1970.

Wilson, E. D., R. T. O'Haire, and F. J. McCrory, comps. *Map and Index of Arizona Mining Districts.* Arizona Bureau of Mines, August 1961.

Witcher, James C. "Thermal Springs of Arizona." *Fieldnotes* 11, no. 2, Arizona Bureau of Geology and Mineral Technology (June 1981).

Newspapers and Periodicals

American Mining Congress Journal

Weekly Arizonian (Tucson, Arizona)

The Arizona Citizen (Tucson, Arizona)

Arizona Daily Star (Tucson, Arizona)

Arizona Miner (Prescott, Arizona). Also, *Arizona Weekly Miner* and *Arizona Daily Miner* as a journal of the Arizona Territorial legislature when it was in session.

Arizona Sentinel (Yuma, Arizona)

Arizona Republican (Phoenix, Arizona)

Congressional Globe

Congressional Record

Daily Alta California (San Francisco, California)

Daily National Democrat (Marysville, California)

Daily Missouri Republican (St. Lewis, Missouri)

Dips, Angles & Spurs: The Newsletter of the Society of Mining Law Antiquarians, www.mininglawhistory.org.

The Elko Independent (Elko, Nevada)

Evening Bulletin (San Francisco, California)

Federal Register

London Mining Journal

Los Angeles Star

Mining and Scientific Press (San Francisco, California)

Mohave County Miner (Kingman, Arizona)

Nevada Democrat (Nevada City, California)

Pittsburgh Post (Pittsburgh, Pennsylvania)

San Francisco Weekly Herald

San Joaquin Republican

Spokesman-Review (Spokane, Washington)

Superior Sun (Superior, Arizona)

The Tucson Daily Citizen (Tucson, Arizona)

The Washington Times (Washington, DC)

Washington Union (Washington, DC)

Weekly Epitaph (Tombstone, Arizona)

Archives and Official Records

Arizona Historical Society, Tucson, Arizona.

Arizona State Archives, Phoenix, Arizona.

Mohave County Historical Society, Kingman, Arizona.

Recorder's Offices:

- Apache County, Arizona.
- Cochise County, Arizona.
- Doña Ana County, New Mexico.
- Greenlee County, Arizona.
- Maricopa County, Arizona.
- Pima County, Arizona.
- Mohave County, Arizona.
- Santa Cruz County, Arizona.
- Yavapai County, Arizona.
- Yuma County, Arizona.

Navajo County, Arizona.
Gila County, Arizona.
Graham County, Arizona.
Pinal County, Arizona.
Coconino County, Arizona.

Mining District Records:
See appendix A and appendix B.

Spanish Archives of New Mexico, Santa Fe, New Mexico.

Special Collections, University of Arizona, Tucson, Arizona.

Index

Page numbers in italics indicate illustrations.

About the Author

John C. Lacy is a mining lawyer and mining law historian who, throughout his career, has analyzed and written about the contributions of mineral adventurers to the development of the law as it was applied to the acquisition of mineral rights. He graduated from the University of Arizona with a degree in journalism in 1964 and one in law in 1967. Lacy has represented mineral clients throughout the western United States and in South America and has taught mineral law and related subjects at the University of Arizona since 1976. His historical activities include work with the Arizona Historical Society, the Arizona History Convention, and the Society of Mining Law Antiquarians.

The subject has come naturally to John as he grew up in the Peruvian mining camp of La Oroya, and during college he worked as a fieldhand for several mining exploration companies. Presently, he is the director of the Global Mining Law Center at the James E. Rogers College of Law at the University of Arizona, where he teaches and supervises several online courses. He also continues to practice law at his firm, DeConcini McDonald Yetwin & Lacy, P.C.

About the Author

[illegible]

Printed and bound by CPI Group (UK) Ltd, Croydon, CR0 4YY
02/09/2026
02208273-0012